The Girmitiya Peasants in Suriname

Ruben Gowricharn

The Girmitiya Peasants in Suriname

Agrarian and Economic Transformations in a Plantation Society

Springer

Ruben Gowricharn
Humanities
Vrije Universiteit Amsterdam
Amsterdam, Noord-Holland, The Netherlands

ISBN 978-3-031-67960-5 ISBN 978-3-031-67961-2 (eBook)
https://doi.org/10.1007/978-3-031-67961-2

Translation from the Dutch language edition: "Miskend Verleden. Hindostaanse boeren in Suriname 1880-1980" by Ruben Gowricharn, © Publisher and Author 2020. Published by Verloren. All Rights Reserved.

This Springer imprint is published by the registered company Springer Nature Switzerland AG
The registered company address is: Gewerbestrasse 11, 6330 Cham, Switzerland

List of Area Measurement Units

are: 10 metres × 10 metres = 100 square metres
hectare, ha: 100 metres × 100 metres = 10,000 square metres
field: 0.05 ha (500 square metres)

Preface

In the collective imagination of Indo-Surinamese, who refer to themselves as Hindostanis, their ancestors were transported from British India to the Surinamese plantations after an exhausting voyage where they were treated and exploited just as in slavery. Nevertheless, they remained in Suriname to make the best of it. And they succeeded because otherwise we, the third, fourth or fifth generation of Hindostanis, would not have existed. The Hindostanis migrated to the city, became proud citizens of Suriname, integrated well and prospered through education and entrepreneurship. Striking in this narrative is the absence of peasants, a segment of the population that was so naturally present in Surinamese society for more than a century. That is my main motive for writing this book.

This book was published in Dutch in 2020 as *Miskend Verleden. Hindostaanse boeren in Suriname 1889–1980*. Hilversum: Verloren van Themaat. Because international colleagues often inquire about a book dealing with Hindostanis in Suriname, and a book about the Hindostani peasantry is lacking, I decided to publish an English version. The Dutch version has been revised and shortened for an English audience. The present book may be considered supplementary to an analysis of the cultural and political history of Hindostanis, *Multiple Homemaking. The Ethnic Condition in Indian Diaspora Societies* (London/New York: Routledge 2020), that deals with different topics concerning homemaking of the Hindostani community.

The book is based on several sources. The most important were the annual reports of the colonial administration, *Surinaamsch Verslag*, and the reports of the Ministry of Agriculture and Economic Affairs (LEZ, whose name changed a few times) and the Landbouwproefstation, a research department of the LEZ. The reports of these agencies were highly irregular. That is reflected in the discontinuity of years or the absence of a publication year. In the text, however, the sources are mentioned as precisely as possible.

The Dutch manuscript was commented upon by quite a few Dutch and Hindostani scholars. Their comments could be divided into those with which I immediately agreed, those that enforced a reformulation and those with which I did not agree at all or for which I offered an explanation. It was striking that there were two dividing lines between the Hindostanis and the native Dutch colleagues. The Hindostanis had no need for details of the lives of individual peasants. In contrast, the native Dutch, with one exception, suggested detailing the hardship of peasants. This difference in preferences can probably be understood from the fact that the Hindostani co-readers came from the world of peasants or were still familiar with it, sometimes remotely. In that case, the desire of the native Dutch readers to specify the individual peasants' life stems from curiosity and a lack of knowledge rather than from an academic need to place the peasants' lifeworld at the centre of this work.

However, I acknowledge that this lifeworld is important in a treatise on peasant history. But although remarks on the lifeworld are scattered throughout the various chapters, I did not find enough material to describe peasant life in detail in the various time periods. Unlike the situation of indentured labourers, who had formal rights and regulators, about whom much was also written, there is little material on peasants. For these reasons, I have used photographs, supplemented by observations, statistics and conclusions from various rapporteurs, to depict peasant life, or aspects of it, in the relevant time period.

The second dividing line between the two categories of co-readers was that most native Dutchmen liked to see a comparison between Hindostani, Javanese and Creole peasants. As tempting as this suggestion is, the question is what such a comparison would yield. It would perhaps satisfy curiosity, but not advance my research goal. Worse, I think such an extension would shift my perspective because it would presuppose a different research design. For this reason, I ignored this suggestion.

In short, although I have been unable to agree with all comments and could not satisfy all my readers, I have taken them seriously by incorporating them or by explicitly explaining my position. The input of the co-readers has improved the manuscript, for which I thank them. My thanks also extend to Thys Verloren van Themaat, the publisher of the Dutch edition, for designing the register of names and subjects.

This publication was supported by a financial contribution from Professor van Winter Fonds.

Rotterdam, The Netherlands
May 2024

Ruben Gowricharn

Contents

Abbreviations

BLO	Bureau for Land Development
CELOS	Centre for Agricultural Research in Suriname
DLS	Third Agricultural Census Suriname
GDP	Gross Domestic Product
HPP	Hindostani Progressive Party
K.v.K	Chamber of Commerce
KTPI	Kaum Tani Persatuan Indonesia
LEZ	Agriculture Economic Affairs
LPS	Agricultural Testing Station
LVV	Agriculture, Livestock and Fisheries
NEI	Netherlands Economic Institute
NOP	Nickerie Independent Party
NPK	National Party Combination
NPS	National Party Suriname
NVNV	National Association 'Nickerie Vooruit'
PSV	Progressive Suriname People's Party
SML	Stichting Machinale Landbouw
TJP	Ten-Year Plan
TLS	Second Agricultural Census Suriname
VHP	United Hindostani Party, since 1973 Progressive Reform Party
VSB	Suriname Business Association

List of Annexes

List of Maps

List and Origin of Photos

List of Tables

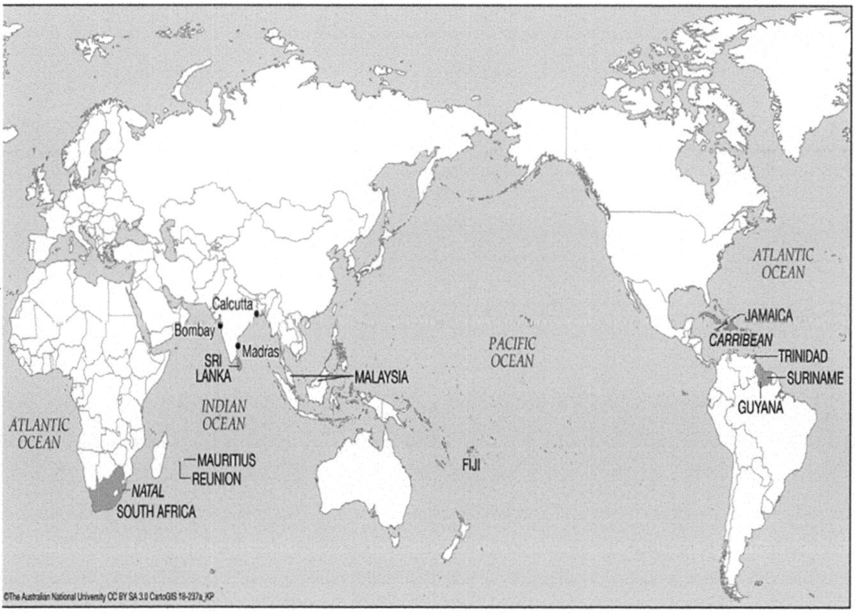

Map 1 The main countries to which British Indian contract workers were shipped. (Source: Lal 2019: 2)

Chapter 1
Peasants in Girmitiya Scholarship

Abstract This chapter argues that Girmitiya scholarship suffers from three blind spots: it has ignored the period when British Indians were primarily peasants. That is remarkable because it was precisely during this phase that their ethnic community formation matured and formed the basis for their further cultural, economic and political development. Second, it has overlooked that the workers experienced three distinct transformations: from various groups directly and indirectly involved in agriculture in British India to plantation workers, from plantation workers to peasants in Suriname, and from peasants to 'urbanites'. Third, Girmitiya scholarship has ignored the sacrifices made by the labourers and peasants. That neglect entail a distorted view of Girmitiya history. The central hypothesis outlined in this chapter is that possessing agricultural knowledge and skills to practise agriculture has determined the three transformations of the British Indians. They could survive and be successful because the agricultural knowledge and skills could be mobilised by the ethnic group.

Keywords Girmitiyas · Peasants · Ethnic community · Transformations · Agricultural knowledge and skills

In the nineteenth and twentieth centuries, British Indian indentured labourers were recruited for work on plantations and sometimes for other work such as the construction of public works in South and East Africa. A characteristic feature of this indentured labourer system was that violations of the contract provisions were punishable by law, the so-called 'Poenal Sanction'. The countries that received the most indentured labourers were Sri Lanka, Guyana, Trinidad, Suriname, Mauritius, Fiji, Reunion, Guadeloupe, Martinique and some smaller territories, and South Africa. Clarke et al. (1990, p. 9) present numbers of emigrants that amount to 1.4 million.

However, Mishra (2015, p. 372), taking into account the number of predominantly South Asian labourers who migrated to Myanmar and Malaya, stated that the number of labourers increased by 6 million. Bates (2017, p. 9) cautioned that the majority of these 'Asian labourers' returned home, although substantial proportions

R. Gowricharn, *The Girmitiya Peasants in Suriname*,
https://doi.org/10.1007/978-3-031-67961-2_1

settled in the societies they visited for work. The Caribbean societies had a reverse experience as the majority settled in the colonies (De Klerk 1953; Laurence 1994).

The fate of these labourers had been discussed by numerous authors (scholars, politicians and officials) but only became widely known with the publication of Hugh Tinker's influential book *A New System of Slavery: The Export of Indian Labour Overseas 1830–1920* (1974). Tinker situated the indentured system in the historical context that was dominated by the abolitionist movement. Every form of bonded labour was related to, or compared with, the recent abolished slavery. It was not surprising that Tinker's central proposition was that slavery was continued with the indentured labour system, albeit in a new form.

Politically and scientifically, the indentured labour system had proponents and opponents as early as the colonial period (for discussions on this, see Eltis 2002; Munro 2014; Northrup 1995). Proponents in this debate argued that while the system of indentured labour had imperfections, the advantages of indentured labour (providing and securing labour for the plantations, thus fixing labour costs) outweighed the disadvantages. Opponents argued that contract labour was a new form of slavery. Since then, research on the emigration of British Indians has focused heavily on the plight of indentured labourers on the plantations. These studies addressed issues such as working conditions on the plantations, exploitation, remuneration, arbitrariness in the distribution of tasks, abuse of women, the health status of the workers, the application of penal provisions and the return to British India.

Some of the studies went beyond immigration and worker life on the plantations and discussed the social and political integration of the indentured into their new country. This is the case, for example, with the studies of De Klerk (1953) and Laurence (1994), by far the most thorough studies of the immigration of British Indian indentured in Suriname and in Trinidad and Guyana, respectively. Nath's (1950/1970) book promises to deal with the history of British Indians and their descendants, but predominantly discusses immigration. The focus on labour immigration hardly differs outside the Caribbean, particularly in Fiji, Mauritius and South Africa (Carter 1995; Freund 1995; Lal 1983, 1992, 2000).

The historical development of the indentured labourers in the various plantation colonies, increasingly referred to as 'Girmityas' ('Girmitya', also written 'Girmitiya', is the Bhojpuri word for people under contract; see Lal 1983, pp. 9–11), is largely similar in the countries mentioned. Nuances aside, it can be summarised as follows: the British Indian indentured labourers were brought into the plantation colonies exhausted, often misled and exploited, but nevertheless managed to survive. After their contract period, they seized the opportunity to settle as part-time or full-time peasants on land provided to them by the plantation owner or the government. Later they moved to the cities, where they received education and excelled in prestigious professions, in politics and in business. In other words, the poor indentured worker from British India had become a successful immigrant. He was celebrated as a hero, described in many volumes, often illustrated with photographs of some prominent descendants. Significant titles such as *From British Indian Emigrant to Citizen of Suriname* (Azimullah et al. 1963) and *Arising from Bondage* (Ramdin 2000), referring to Trinidad and Guyana, are telling.

While scholarly attention has largely focused on the lives of indentured labourers, there has been a staggering lack of interest in the period when British Indians were primarily agriculturalists. That period lasted nearly a century and included several generations, the first of which consisted of peasants who used to be plantation labourers. The lack of attention given to British Indian peasants is remarkable because it was precisely during this phase that their ethnic community formation matured and formed the basis for their further cultural, economic and political development. This lack of scholarly attention to the British Indian peasants and their agrarian descendants may qualify as a blind spot in the historiography of Hindostani peasants.

The Girmitiya scholarship harbours a second blind spot. It has overlooked the fact that the workers experienced three distinct transformations. The first was the transformation of various groups directly and indirectly involved in agriculture in British India to plantation workers, the second was from plantation workers to peasants in Suriname, and the third was from farmers to 'urbanites'. The individuals involved in the first transformation consisted of peasants, land labourers, artisans and other occupational groups embedded in agriculture (De Klerk 1953, pp. 98–112). By far the majority of studies have addressed only the first transformation, specifically recruitment in British India and life on the plantations in the various overseas colonies. This book broadens the focus to three transformations. By connecting the three transformations, this book purports to make clear that there is an economic and sociocultural continuity between the origin of the indentured labourers, the establishment as peasants in the colonies and the transformation to urbanites.

A third blind spot consists of ignoring the sacrifices made by the indentured labourers. This 'price of progress' is not limited to the plantation workers but includes the first generation of peasants who were previously labourers. Since this generation, whose settlement was substantial, several generations have populated the agricultural sector in Suriname. Because life spans before and at the beginning of the twentieth century were short and health was poor, peasants died earlier. The price they paid consisted not only of lost lives but also their long-term suffering. Unfortunately, this price hardly comes into the picture except as juicy narration or in music, and then in a limited form, because existing research focuses mainly on progress (Choenni 2016, pp. 646–658).

This book aims to analyse the history of the Girmitiya peasants in Suriname, a former Dutch colony that harbours descendants of British Indian indentured labourers. It uses the term 'British Indian peasants' because the first generation of immigrants were largely British subjects. Even in the Surinamese literature on agriculture, they were mainly referred to as 'British Indians'. In 1927, these immigrants and their descendants received Dutch citizenship. From that period, the term 'Dutch' or 'Surinamese peasants' (because British India is a country like the Netherlands and Suriname) could be used, but they will be referred to as 'Hindostani' or 'Girmitiya peasants'. The first label is because members of the ethnic community called themselves 'Hindostanis', while the second label is current in Girmitiya scholarship.

To survive on plantations and as peasants, knowledge is required of the agricultural process. Agrarian knowledge is embodied in peasants and consists of knowledge of the soil and climate, as well as cultivation, conservation, processing, preparation and consumption of the product. The concept of human capital fits the requirement of knowledge. It refers to individuals' investment in education through which they increase labour productivity in order to earn a steady or higher income in the future (Becker 1964). As this knowledge relates to the agricultural sector, I call it 'agricultural human capital' (Gowricharn 2020).

Agricultural human capital is not an individual but a collective asset of the farming community. Human capital in its modern usage is acquired individually through education and training, while agrarian human capital is acquired through traditional transfer (a kind of 'training on the job'), habit and routine in the lived world. Therefore, agricultural human capital can also be referred to as 'collective' or 'communal' capital. Cooperation is characteristic of agricultural human capital. It takes shape through traditions and collective habits and is often decisive in sustaining itself, especially in peak periods of planting and harvesting, as has often been observed.

The Girmitiya peasants became part of an ethnic community. Most definitions of ethnic groups go back to Max Weber (1978, p. 389), who defined ethnic groups as people that entertain a subjective belief in their common descent, be it a physical type, customs or both, or because of memories of colonisation and migration. The belief must be important for group formation, whereas it does not matter whether an objective blood relationship exists. Unlike professionals and sportsmen, people are born into an ethnic group, grow up in it and acquire a collective and individual cultural identity (Hale 2004). In most Caribbean countries, as well as in plural societies like Singapore, ethnicity and race coincide. Despite his opinion that ethnic groups will become dissolved in the modernisation process, Max Weber (1978) believed that an ethnic bond is something like a family because of the emotional attachment to the group. He used the term 'family resemblance'. The notion of a family resemblance was developed by Edward Shils (1957) into the term 'primordial bond' or 'ties'. Later, Geertz (1971) further developed this concept into 'primordial loyalties'. He argued that primordial loyalties were rooted in race, creed, region, language, religion and the like.

The recruited British Indians did not yet constitute an ethnic group in India. They did possess some commonalities such as appearance, country of origin and arrival and the like, but there were regional differences in terms of language, religion, rituals, morals and customs. Mayer (1966, p. 97) called such a group an 'interactive quasi-group', which he described as a collection of people that 'possesses a degree of organisation, but is nevertheless not a group'. The group is thus yet to emerge.

Ethnicity also comprises productive forces; examples include the Protestant work ethic addressed by Weber (1904), arguing that a strong work ethos combined with thrift explained the rise of capitalism. Another 'interconnectedness' between economics and culture (or ethnicity) can be found in what is referred to as 'local knowledge' (Geertz 1983, p. VIII): this is knowledge that cannot be generalised, for example because the knowledge is specific—think of weather conditions, types of soil, agricultural practices, cultivation and the use of traditional medicine and powerful herbs. When local knowledge is embodied and peasants are bearers of that knowledge, it becomes agrarian human capital. Hence, the knowledge (including the skill to operate labour instruments) belongs to the productive forces and is part of the ethnic community. The ethnic bonding, representing a moral community, also accounts for cultural incentives to mobilise resources (Carrier 2018), including labour, seeds, instruments, draft animals and money.

The productive forces, reflecting the economic capability, were part of the culture of the peasants. These elements of culture and economy are essential to hold your own in society. What is special about this perspective is that the production capacity of the peasants was a collective force, not an individual one as is often postulated in conventional economics (Udehn 2002). So by culture here is meant not the arts and literary expressions but all the knowledge, routines and skills that make production possible. Culture is by definition a collective asset, and in this book it is relevant only when it is intertwined with the economy. This is not to say that other dimensions of culture are not relevant, but in this work they are of little significance.

The absence of scholarly attention to the peasants is reflected in the three blind spots mentioned earlier. These are: the disregard for the origins of the ethnic group and its connection to the productive capacity of the peasants; the disregard for the three transformations that British Indian peasants went through historically, so that continuity and ruptures are not adequately portrayed; and the disregard for the price that peasants and their descendants paid to give later generations a better future. This book thus emerges from a composite problem in the historiography of Hindostanis and cannot be captured in a central research question.

On the basis of the problems outlined above, I arrive at three objectives:

1. To explain how ethnic group formation was related to the economic position of British Indians and thus determined the productive capacity of the peasants.
2. To make an analysis of the three transformations to show the historical changes and thus the continuity and fractures.
3. To detail the sacrifices the Girmitiyas—labourers, peasants and their descendants—made.

The working hypothesis of this book is that possession of the agricultural knowledge and skills to practise agriculture determined the three transformations of the British Indians. This economic capability could only be mobilised by an ethnic group because the productive forces were contained within it. Thus, in this perspective, ethnic group formation is the independent variable and economic performance is the dependent variable. The distinction is analytical because the productive forces were inextricably intertwined with culture. However, ethnic group formation was not a simple reproduction of the culture, as it also included the modified ethnic institutions and group culture. These are small and large adjustments that are typical of migrants when they try to build a new life as a group in a foreign country (Alexander et al. 2018; Gowricharn 2013).

What does the working hypothesis explain? First, that there is a link between group culture and the agrarian economy. The British Indians harnessed productive forces that were inherent in group culture. Therefore, these forces could only be mobilised as a group. Second, that the recruitment of the labour force for Suriname was motivated not only by the substitution of slave labour, as several authors have argued (Bhagwanbali 1996; De Klerk 1953; Snellen 1933), but also by the agricultural human capital of the workers. Third, that the three transformations were not isolated phenomena but had a continuous line of immigration, colonisation and urbanisation. The first two transformations were made possible by agrarian productive forces, while the third originated because this capital became obsolete. Thus, all three transformations had more or less the same causal forces.

Theoretically, the three transformations are approached through three different conceptual lenses. The first transformation to workers is known as 'proletarianisation'. It is a process that Karl Marx (1867) referred to in the first part of *Capital* I as a form of 'primitive accumulation'. In Marx's depiction, peasants in seventeenth-century England were violently separated from their land. That generated two effects. First, the expropriation of peasants' land led to small pieces of land being merged into larger plots and exploited by an agrarian capitalist class. The second effect was the emergence of a class of people without land, who were thus forced to work elsewhere. The job seekers—in fact, the labouring class—were absorbed into the then-expanding industrialisation. Proletarianisation does not need to be accompanied by physical violence, however. When peasants suffer misfortune because the harvest fails, life becomes more expensive; or because their land may be too small to feed all the family members, the agrarian enterprise must close and members of the peasant family are forced to work wholly or partially as wage labourers.

The second transformation, from workers to peasants, is known as 'peasantisation'. The English-language literature distinguishes between peasants and farmers (Goodman and Redclift 1981). Generally, the former term refers to small agriculturists who produce mainly for their own consumption, but recent scholarship discloses a wide variety of peasants in disparate historical contexts (Service, year unknown). Farmers, on the other hand, are agriculturalists who produce for the market, specialise in a limited number of products and are largely incorporated into the cash economy. The distinction between farmers and peasants is typological; in practice, they are hard to find in 'pure' form. This is mainly because today's economies are highly monetised, requiring every peasant to sell in order to buy other products with the money earned.

The third transformation concerns the migration of peasants to the city. This transformation has been captured by the concept of mobilities (Schiller and Salazar 2013; Sheller 2014). The concept does not refer to social rise or fall as in classical sociology, but the coherence between various changes. These include the changing relationship between mobility and immobility (i.e. the left behinds), between local and transnational connections, between being rooted in a particular area and developing a cosmopolitan orientation. Mobilities also include the various transformations as they occur in migration: the spatial mobility from one place to another, the social ascent from one environment to another, the economic change of occupations, the political shifts, and the many large and small cultural changes. The disadvantage of this concept is that it puts very different changes under one umbrella and tries to 'capture' a cocktail of phenomena (Sheller 2014). The advantage, however, is that the various changes can be viewed in conjunction with each other.

A brief word on the choice of Suriname. The transformation from worker to peasant has not been completed in all plantation economies. In Jamaica, for example, where land was scarce because the planters had a claim on it or because it was not suitable for farming, British Indians were largely assimilated (Shepherd 1993). In most other countries, the land was there but access was intentionally kept restricted to force workers to work as part-timers on the plantations (Allen 1999; Gowricharn 2013; Lal 2000). Of all the countries mentioned, the transformation from labourer to farmer was most complete in Suriname. This was because the plantation sector hardly existed after the 1920s. As a result, land was abundantly available and the government was able to encourage the settlement of immigrants. Because Suriname had the most complete development of the peasantry, it is therefore the most suitable object for studying its transformations.

1.1 The Structure of This Book

The periodisation begins in 1880 when the first batch of Girmitiyas had served their contract and were able to settle themselves as peasants. For several reasons, the peasants were marginalised and moved to the city, where they were employed in the public sector. Around 1980, the Hindostani peasant ceased to exist. Based on these considerations, I mark the period to be studied as from 1880 to 1980.

The theoretical concepts aim to capture the basic theoretical relation between ethnicity and productive forces. The importance of theoretical concepts is that data can be framed and understood, that the concepts point to specific developments and that comparisons can be made with corresponding developments elsewhere. The theoretical concepts in this book have primarily a pointing function, i.e. they are a lens, and they are not intended to produce new theoretical insights but rather point at a direction to look at as Blumer (1954) has suggested. The use of theoretical concepts also prevents the reader from being overwhelmed by a lot of data that, although interesting, do not lead anywhere.

Each chapter is historically delineated, discusses a specific topic and has a separate framework for analysing and interpreting developments. That framework is found in the chapter's introduction and is evaluated each time in the conclusion of the same chapter. Thus, each chapter can also be read separately.

The book is divided into three parts. Part 1 covers the period 1873–1920 in which the supply of labourers and most of the returns took place. It is also the period when the community building emerged and workers maintained and updated their agricultural knowledge and skills. These developments are described in three chapters. The central topics in this part are the conditions under which the workers were recruited in British India, shipped to Suriname and put to work on the plantations, and how, in this process, the first beginnings of ethnic communalisation arose (Chap. 2). The experiences on the plantations, measured by the harshness of the labour process, the earning capacity and the amount of savings, were strongly determined by physical conditions, the regime on the plantations and family formation (Chap. 3). Through various adaptations, a plantation culture emerged that would characterise the British Indian community. Out of the many ways to depict a culture, I chose the reproduction of rituals in the lived culture, the erosion of the caste system and gender inequality, expressions of emotions and interracial relations (Chap. 4). The chapters of this first part analyse the initial transformation from agriculturalists in British India to wage labourers in Suriname as a cultural and economic process. The central point in this part is that the workers began to develop an ethnic community and managed to retain their agricultural knowledge and skills. These two realities were closely intertwined from the beginning.

The second part covers the period from 1895, the beginning of the colonisation policy, to 1950 when rice production was modernised. Central topics include the emergence of the preconditions for peasant settlement, particularly the closure of plantations that set labour and land free and enabled the rise of a small peasant sector (Chap. 5). Contrary to a strong current in the Girmitiya literature about plantation labourers, the experiences on the plantation were an important push factor toward settling outside. That process went hand in hand with making a home by expanding the ethnic group, rearranging the agricultural landscape and developing a rice economy. This part also explores the connection between the productive resources of the ethnic group—particularly the use of agricultural knowledge that was available in the community and cooperation among peasants—and economic performance (Chap. 6). Moreover, these developments did not prevent increasing numbers of British Indians from seeking refuge outside agriculture. This professional differentiation was observable in both districts and the cities (Chap. 7).

Part three covers the period 1950–1980. It discusses the forces that undermined the hitherto productive link between ethnic community and rice economy. These consisted mainly of persistent poverty, mechanised rice farming, Western-oriented education and migration to the city. As mechanised rice farming did not lead to financial independence for Hindostani peasants, their migration from the districts to the city continued (Chap. 8). The weakening of ethnic cohesion was hardly addressed by the Verenigde Hindostaanse Partij (VHP, United Hindostani Party) that represented the community. This was partly because mechanised agriculture had been conceived by the Netherlands and partly because the VHP was led by an urban-oriented elite. However, many district residents practised agriculture as a secondary activity, held white-collar jobs and developed an urban lifestyle (Chap. 9). The peasantry shrank even more rapidly on the eve of independence in 1975 when many half and whole rice producers left with their families for the Netherlands. This development did not stop until 1980 when family reunification came to an end (Chap. 10). The concluding chapter evaluates the objectives of this book by highlighting the main findings.

References

Alexander C, Chatterji J, Jalais A (2018) The Bengal diaspora. Rethinking Muslim migration. Routledge, London

Allen R (1999) Slaves, freedman, and indentured laborers in colonial Mauritius. Cambridge University Press, Cambridge

Azimullah E, Ganpat H, Lutchman W (1963) Van Brits-Indisch Emigrant tot Burger van Suriname. Wieringa, s-Gravenhage

Bates C (2017) Some thoughts on the representation and misrepresentation of the colonial south Asian labour diaspora. South Asian Stud 33(1):7–22. https://doi.org/10.1080/02666030.2017.1300372

Becker G (1964) Human capital: A theoretical and empirical analysis with special reference to education. Chicago University Press, Chicago

Bhagwanbali R (1996) Contracten voor Suriname. Arbeidsmigratie vanuit Brits-Indië onder het indentured-labourstelsel, 1873–1916. Doctoral dissertation, Radboud University (Nijmegen)

Blumer H (1954) What is wrong with social theory? Am Sociol Rev 19(1):3–10

Carrier J (2018) Moral economy. 'What's in a name? Anthropological Theory 18(1):18–35

Carter M (1995) Servants, sirdars and settlers. Indians in Mauritius 1834–1874. Oxford University Press, Delhi

Choenni C (2016) Hindostaanse contractarbeiders 1873–1920. Van India naar de plantages in Suriname. LM Publishers, Volendam

Clarke C, Peach C, Vertovec S (eds) (1990) South Asians overseas: Migration and ethnicity. Cambridge University Press, Cambridge

De Klerk CJM (1953) De Immigratie der Hindostanen. Urbi et Orbi, Amsterdam

Eltis D (ed) (2002) Coerced and free migration. Global perspectives. Stanford University Press, Stanford

Freund B (1995) Insiders and outsiders. The Indian working class of Durban 1910–1990. Heinemann, Portsmouth NH

Geertz C (1971) The integrative revolution: Primordial sentiments and civil politics in the new states. In: Welch C (ed) Political modernization: A reader in comparative political change Belmont. Duxbury Press, CA, pp 198–218

Geertz C (1983/2000) Local knowledge: Further essays in interpretative anthropology. Basic Books, New York

Goodman D, Redclift M (1981) From peasant to proletarians: Capitalist development and agrarian transitions. Basil Blackwell, Oxford

Gowricharn RS (2013) Ethnogenesis: The case of British Indians in the Caribbean. Comp Stud Soc Hist 55(2):388–418

Gowricharn RS (2020) Multiple homemaking. The ethnic condition in the Indian diaspora. Routledge, London

Hale H (2004) Explaining ethnicity. Comp Pol Stud 37(4):458–485

Lal B (1983) Girmitiyas. The origins of the Fiji Indians. Australian National University Press, Canberra

Lal B (1992) Broken waves. A history of the Fijian Islands in the twentieth century. University of Hawaii Press, Honolulu

Lal B (2000) Chalo Jahaji. On a journey through indenture in Fiji. Australian National University Press, Canberra

Laurence K (1994) A question of labour: Indentured immigration into Trinidad and British Guiana 1875–1917. Ian Randle Publishers, Kingston

Marx K (1867) Capital. A critique of political economy, vol 1. Penguin Books, Middlesex

Mayer A (1966) Quasi-groups in the study of complex societies. In: Banton M (ed) The social anthropology of complex societies. Tavistock Publications, London, pp 97–122

Mishra AK (2015) Sardars, Kanganies and Maistries: Intermediaries in the Indian Labour Diaspora during the Colonial Period. In: Wadauer S, Buchner TH, Mejstrik A (eds) The History

of Labour Intermediation: Institutions and Finding Employment in the Nineteenth and Early Twentieth Centuries. Berghahn Books, New York, Oxford, pp 368–387. https://doi.org/10.1515/9781782385516-014
Munro D (2014) Conclusion: On resistance and accommodation. In: Lal B, Munro D (eds) Hassankhan M. Resistance and Indian experiences. Comparative perspectives, Manhor, pp 293–311
Nath D (1950/1970) A history of Indians in Guyana. Second revised edition published by the author, London
Northrup D (1995) Indentured labor in the age of Imperialism, 1834–1922. Cambridge University Press, Cambridge
Ramdin R (2000) Arising from bondage: A history of Indo-Caribbean people. Tauris Publishers, London/ New York
Schiller N, Salazar N (2013) Regimes of mobility across the globe. J Ethn Migr Stud 39(2):183–200
Sheller M (2014) The new mobility paradigm for a live sociology. Current Sociology Review 63(6):789–811
Shepherd V (1993) Transients to settlers: The experience of Indians in Jamaica 1845–1950. Peepal Tree Books, Leeds
Shils E (1957) Primordial, personal, sacred and civil ties. Br J Sociol 8(2):130–145
Snellen E (1933) De Aanvoer van arbeiders voor den landbouw in Suriname. Doctoral dissertation, University of Wageningen
Tinker H (1974) A new system of slavery: The export of Indian labour overseas 1830–1920. Oxford University Press, London/New York
Udehn L (2002) The changing face of methodological individualism. Annu Rev Sociol 28:479–507
Weber M (1904) The Protestant ethic and the spirit of capitalism. Allen & Unwin, London
Weber M (1978) In: Roth G, Wittich C (eds) Economy and society: An outline of interpretative sociology, vol 2. University of California Press, Berkeley

Part I
From Rural Dweller to Plantation Labourer

Chapter 2
Migration and Agrarian Human Capital

Abstract This chapter discusses the features of Girmitiyas that enabled their survival as labourers and peasants and establishing an ethnic community. It starts with a reflection on the historical context of labour migration, which was characterised by British Indian agriculture burdened by the advance of the money economy, limited land tenure and landlessness, and periodic droughts. This led to a high propensity to migrate and the emergence of a migration culture. Sect. 2.3 looks at the agricultural human capital of workers. This is a little-noticed feature of labour demand. Recruitment, especially recruitment practices and procedures, is examined in Sect. 2.4. Section 2.5 discusses the journey to Suriname, the route along which the *dipuá* and *jaháji* relations came into being. Section 2.6 zooms in on the demographic characteristics of the immigrants, specifically cast and male-female disparities, which is important to put the experiences on the plantations into perspective.

Keywords Agriculture · Migration culture · Recruitment · Journey · *dipuá* and *jaháji* relations · Girmitiya demographics

2.1 Introduction

The transformation from rural people to labourers on plantations in Suriname consists chronologically of three phases, namely the recruitment, the subsequent stay in the depots in Calcutta and the shipment to the country of destination. Discussion of these phases mainly revolves around the push factors, the misleading recruitment and the suffering of the workers. This inevitably limits the perspective to individual fates and anecdotes. Authors who emphasise individual fates (Bhagwanbali 2010; Choenni 2016; De Klerk 1953) overlook the encompassing forces that led to proletarianisation. They are best understood from the workers' agrarian background and the migratory culture that existed at the time. To make sense of this process of proletarianisation, I consider the historical context of the British Indian peasants (Map 2.1).

R. Gowricharn, *The Girmitiya Peasants in Suriname*,
https://doi.org/10.1007/978-3-031-67961-2_2

The historical picture of the indentured labourers is heavily shaped by recruitment. *A New System of Slavery* (Tinker 1974) cannot, of course, be based on free will emigration—the title suggests that recruitment, deception and/or violence were involved as happened with the capture of slaves. Deception by recruiters is not ruled out, but it is miraculous that this element in the recruitment story is still dominant despite recent studies reporting a culture of migration (Kumar 2017). Moreover, emigrants have been wrongly presented as unskilled and undifferentiated labour, as 'muscles'. Their possession of agricultural human capital has been overlooked by virtually all previous authors about the Girmitiya diaspora (Carter 1995; Choenni 2016; De Klerk 1953; Gillion 1962; Laurence 1994), but agricultural human capital is important for understanding economic labour demand and peasant performance.

Bringing together people with the same human capital already creates a certain degree of kinship. This is a well-known phenomenon in the history of labour migrants (such as guest workers in Europe or immigrants in the US), but kinship leads more easily to a relatively uniform community when the workers are somewhat familiar with each other's lived cultures. Fenton (2003, p. 115) calls this 'tacit ethnicity', which can be better referred to as 'latent ethnicity' as it includes the potential to become an ethnic property. Latent ethnicity manifested itself among British Indians in marriages concluded in depots and in assumed relationships, so-called *dipuá* and *jaháji* relationships (Gillion 1962). The effects of latent ethnicity are discussed in detail in the next chapter; here I suggest that the development of ethnic relationships was already taking place in the phase of selection and transportation, a phenomenon reported by many authors (Choenni 2016; Lal 1983a; Tinker 1974).

This chapter discusses the migration culture, agricultural human capital, recruitment, crossing and demographic characteristics of the workers. These issues are important for further specifying the character of labour migration from Uttar Pradesh and Bihar, determining the agrarian human capital of the peasants, and interpreting the recruitment and selection and the concomitant emergence of new social ties among the migrants. Following authors such as Lal (1983b) and Roopnarine (2003), I argue that labour migration was triggered by British capitalism, especially the introduction of the money economy, and to a lesser degree by periodic droughts. Proletarianisation was sometimes full and sometimes incomplete, partly permanent and partly part-time, and developed into a tradition of migration. In addition, new social relations, *dipuá* and *jaháji* relations, emerged in the process of recruitment and selection, which later proved significant in the formation of ethnic communities.

In the next section, I reflect on the historical context at the time of labour migration, which was characterised by an agriculture burdened by the advance of the money economy, limited land tenure and landlessness, and periodic droughts. This led to a high propensity to migrate and the emergence of a migration culture. Section 2.3 looks at the agricultural human capital of workers. This is a little-noticed feature of labour demand. Recruitment, especially recruitment practices and procedures, is examined in Sect. 2.4. Section 2.5 discusses the journey to Suriname, the route along which the *jaháji* relations came into being. Section 2.6 zooms in on the demographic characteristics of the immigrants, which is important to put the experiences on the plantations into perspective. The chapter concludes with the main findings.

2.2 The Migration Culture

Most migrants to Suriname were from the United Provinces, Oudh and Bihar areas, which together provided 93% of the workers (Bhagwanbali 1996, p. 74). The naming of the United Provinces changed over the years. De Klerk (1953, p. 47, notes 5 and 6) mentions that the United Provinces were denoted North Western Provinces, in 1902 renamed as United Provinces of Agra and Oudh, and from 1937 onwards they were known as United Provinces, abbreviated to UP. Since 1950, UP has stood for Uttar Pradesh. Bihar belonged to Bengal at that time and had been a separate province since 1912. These provinces in turn consisted of districts, all of which had agriculture as their main means of livelihood (Lal 1983a, Ch. III).

Peasants were slowly disengaged from the land (De Haan 2002; Lal 1983a; Yang 1979), with two processes being decisive. The first was the rise of the money economy. In the late eighteenth century, English administrators introduced the provision to pay taxes in cash (especially to finance their armies), which was followed by creditors and by large landowners who had leased land to peasants. The earlier payment in kind by peasants to large landowners, governments and creditors thus gave way to payment in cash. This need for money, which incidentally was always there to a limited extent, forced peasants and land workers to go out to work during low-season periods. In cases where they did not manage to earn enough money or did not produce enough, they lost the land.

Droughts and rains constituted temporary shocks in peasants' lives and not a permanent erosion of livelihoods like proletarianisation, which profoundly changed livelihood modes and social relations. Roopnarine (2003) points out that the push and pull model falls short in explaining migration because it ignores the underlying processes of dislocation. In this regard, Kumar (2017, p. 43) notes: 'Experiences of extreme hardship, including poverty, famine, and caste-based oppression, formed some decision to migrate within a context of structural coercion, making the suggestion of a clear-cut "agency" or "free will" problematic.' That is to say, local society was most affected by the expansion of capitalism, which led to a high migration propensity.

In addition to the slow transformation of agriculture in the north-east of British India, the type of agriculture was also important for migration. One of the few authors to shed light on this is Bisnauth (2000, pp. 22–23). He states that in key districts from which indentured labourers came, such as Basti, rice could be planted three times a year. In all districts, peasants also cultivated maize, wheat, millet, sugarcane, tobacco, spices and vegetables. The agricultural annual calendar, which was determined by rain (monsoons) and drought, was such that there were intermediate periods of growth and ripening. During those less busy periods, peasants and agricultural labourers preferred to use the time to earn money elsewhere. To do so, they moved to more urban areas. Yang (1979) aptly characterised this behaviour as 'peasants on the move'.

Thus, the job seekers were mostly peasants and agricultural workers who were staying in urban areas when they were recruited. But the recruited population was more heterogeneous: it included artisans and many other occupational groups in the agricultural areas besides peasants and agricultural workers. These non-agricultural occupational groups were part of the agrarian milieu because they were engaged in part-time farming. Thus, the occupations were not completely segregated (Roy 2018). I conveniently refer to all these occupational groups as 'rural' because they were more or less embedded and involved in agriculture.

A similar description is also employed by Alexander et al. (2018, Ch. I), who point out that the domestic migrants consisted mainly of former peasants, agricultural workers and artisans who were equally involved in agriculture. In these circumstances, it is more than likely that most 'city dwellers' had an agricultural past and even practised agriculture part-time. From this population, workers destined for overseas plantations (and tea plantations in Assam) were recruited. Without suggesting that every recruited labourer was an agrarian, it is safe to assume that the lion's share of them had, albeit to varying degrees, agricultural human capital.

Bisnauth (2000, pp. 24–25) describes the location of recruitment as follows: '[W]hile actual recruitment took place in the "towns", the recruits themselves were from the villages ... it must be borne in mind that an Indian "town", particularly one "up country", was nothing more than a large village with a magistrate and a petty court to manage its judicial affairs.' This description of the 'town' has major consequences for the definition of the workers' background. They were peasants, at least agriculturists, even if they lived in the 'city'. For what was defined as 'city' in the statistics was mostly agricultural area, especially when the transition from village to urban area was smooth. Hira (2000, p. 37) estimates that of the more than 34,000 people who arrived in Suriname, about 17,000 were from villages. The rest should then have come from 'cities' that were embedded in, and part of, the agrarian environment.

A persistent misconception in the literature on migration is that the British Indian peasants would not have been very willing to migrate. For example, De Haan (2002) notes that British colonial officials more often than not debunked the view that the population had little mobility. De Klerk (1953, p. 57) writes that the masses of India 'would rather suffer and die on Indian soil than leave overseas for a foreign country'. This view is also echoed by Bhagwanbali (2010, p. 49), who says there was a 'low willingness of the Indian people to emigrate'. The consequence of this viewpoint is that explanations for emigration mainly highlight push factors: poverty, droughts, low land ownership and a hopeless existence. In this perspective, the migrants were individual victims of circumstances and lacked agency.

There are some criticisms of this explanation. The agricultural cycles already indicate a pattern of seasonal migration, although it can be assumed that it was less compelling in the preceding period when the cash economy was less developed. But the main objection to the link between, for example, drought and migration is the implication that emigration started only when droughts occurred or that migration declined during periods when there was no drought. However, historical research on migration suggests otherwise. For Bihar, De Haan (2002, pp. 122–123) points to a flow of migration dating back to the Moghul period. According to his sources, the South Asian rural population has been highly mobile, reflected in large groups of people who moved over large distances. He reports that in the early-colonial period, population movement was the rule rather than the exception.

This migration was so normal that later authors characterised it as a 'culture of migration' (Alexander et al. 2018; Kumar 2017; Narayan 2017). In the period starting from 1929 and into the 1930s, the compilers of The Report of the United Provinces Banking Enquiry Committee noted: 'Twenty years ago it was commonly said that there was not a single family in Banaras division which did not have at least one member in the provinces of Bengal, Assam and other states' (quoted in Majumder 2010, p. 57). This estimation concerns only domestic migration. Majumder (2010) notes that mobility was initially limited by the low-level development of railways. Emigration accelerated when railways were developed and steamships were introduced. Short-term migration to cities also increased, a development the author refers to as the 'Bhaiyá express'. Alexander et al. (2018) also point to this long-standing mobility, distinguishing between short- and long-term migration in the British colonial empire and foreign migration. They estimate the size of domestic migration in the second half of the nineteenth century to be around 30 million people (Alexander et al. 2018, p. 43).

The migration culture further consisted of several salient elements, the most important of which were monetary earnings. Also typical of that culture were the stories and songs that emerged in the region and still persist, stories about the departure and sometimes migration of relatives and the grief of those left behind (Hassankhan 2013; Narayan 2017). Other authors report that the knowledge of migration, recruitment and related procedures was so widespread among the population that some people pretended to be potential migrants, took advantage of all the facilities offered by the recruitment system, such as medical treatment, free trips to the big city, food, stays in the depots and so on, and then left the compound (Kumar 2017, pp. 61–62). These phenomena not only contradict the image of deception but also indicate the existence of a mobile population that was familiar with the recruitment practices, procedures and extant rights.

Job seekers knew about the existence of foreign labour markets. According to the Grierson report (1883), migrant labourers had separate names for various foreign countries: Mauritius was called 'Mirich', Trinidad was known as 'Chinitat' and Demerara was referred to as 'Damra'. Migrant workers in Suriname also knew these names (De Klerk 1953, p. 61). The possession of this knowledge reflects the familiarity with foreign labour markets. Labour migration abroad was not always a one-time occurrence. Many migrants had themselves been recruited several times as indentured labourers. Sometimes they returned to India first and sometimes they left for other destinations on their own initiative after completing their labour contract (Kumar 2017, Ch. II). It cannot be ruled out that the workers sometimes had overblown expectations or that their knowledge of the destination was vague and, as a result, disappointment was around the corner. But such experiences are inherent in migration, both domestic and foreign.

Emmer (1985, pp. 150–151), relying on a report by the Surinamese emigration agent around 1901, points out that among the recruited workers in the depots, there were often return migrants waiting to leave for the colony. He quotes from an official report of the Surinamese emigration agent, who writes: '[H]e (the return migrant, RG) mixes daily with a large number, sometimes as many as 800, of his colleagues. They freely discuss among themselves the situation they are in; they have the benefit of the advice rendered them by a few return migrants who are embarking on a second turn of indenture, and of whom there are invariably a small number present. So it may be taken for granted that at the time of embarkation, 99% of the adult emigrants are fully aware of the kind of life that awaits them on the other side and also that any emigrant who may originally have been recruited against their will has become fully reconciled to heir fate, after discussing the matter in all its bearings with others in the depots' (Emmer 1985, p. 151).

Various stories about these attractive destinations circulated, especially about high wages, great security, good treatment and sometimes the short travel distance. Some authors share the view that migrants mostly knew where they were going: 'The indentured labourers were simple people from simple backgrounds, but they were not simpletons,' wrote Brij Lal (2006, p. 371). In contrast, Emmer wrote, even more resolutely (1986, p. 193): 'There is not a single indication that any of the colonial recruitment agencies were interested in the kidnapping of emigrants or in using fraud and deception as a regular practice.'

However, some authors underline stories of deception by recruiters (Bhagwanbali 2010, p. 60ff; Choenni 2016, pp. 116–126; De Klerk 1953, pp. 55–57). In their representation, the migrants were promised a land of milk and honey and it was suggested that they would be transported to the land of Sri Ram (the Hindu god), which bears some resemblance in sound to 'Sarnam' (as Suriname was referred to). This viewpoint is difficult to reconcile with the knowledge present among the workers of the plantation colonies.

De Klerk (1953, p. 62ff) recounts that the recruited migrants were taken to the subdepots, where they were provided with food and cooking utensils free of charge and where it was 'instilled' in them that they had to declare that they were leaving of their own free will. After a medical examination, they were taken to the administration office for registration. Here, they entered into a preliminary contract with the recruiters. They were then taken to the main depot in Calcutta where they underwent another medical examination focused on their suitability for field and factory work, and they had to declare again before the Protector of Emigrants and the emigration agent that they were leaving of their own free will and had been adequately briefed on the travel destination (see Choenni 2016, Ch. V, for a detailed description). Only then was a final work contract concluded. Thereafter, they stayed in the main depot in Calcutta waiting for a sufficient number of migrants for the ship to depart. Before departure, they had to declare once again that they were leaving of their own free will (see also Emmer 1985).

Since labour migration from British India was more or less normal, the experienced migrants had knowledge of the destinations and work. Depending on the experience of the individual labour migrant, this knowledge was sometimes vague and sometimes accurate. Households in the area of origin received remittances from the migrants. But upon return, the reintegration of some migrants was disappointing. Some people had lost their original status and sometimes their possessions (Choenni 2016; De Klerk 1953; Kumar 2017). This disappointment led to renewed, and often permanent, departure for the outside country.

2.3 Agricultural Human Capital

Carney (2001, p. 151ff) describes a situation where rice planters in South Carolina asked slave traders for slaves who were familiar with rice culture. While this was a different rice culture, with dry rice farming and other rice varieties, it was notable that the demand for slaves was partly determined by their knowledge of the cultivation and consumption of this crop. The possession of agricultural knowledge, of agricultural human capital, was thus a determinant of labour demand. Analogously, the question can be raised as to whether planters sought workers from the northeastern part of British India because they possessed knowledge of growing, in particular, sugarcane, which was the main product in the British colonies (Northrup 1995, p. 29ff).

The Surinamese literature on the indentured labourers hardly addressed this knowledge element of crop cultivation. All attention was on the health of the workers, their suitability for fieldwork and provisioning (Bhagwanbali 2010, p. 25ff; Choenni 2016, p. 41ff; De Klerk 1953, p. 31ff). The focus on suitability for fieldwork reduced the assessment of workers to possessors of muscle power. That is understandable from the historical frame of reference consisting of slavery (Northrup 1995). The indentured labourers were seen simply as substitutes for slaves who were previously also approached as 'unskilled labour'. But research suggests that slaves also possessed human capital and were familiar with the cultures of their masters (Price 2007, p. 22ff).

The link between peasants' expertise and the demand for labourers has been argued in the Indian literature on foreign labour migration. Some mention this determinant of the labour demand in passing (Yang 1979) while others are quite specific (Kumar 2017, p. 155). The most explicit author on the relationship between the agricultural human capital of the indentured labourers and the demand for British Indian labour is Majumber (2010, p. 17), who asserted that the labourers who engaged in cane cultivation in the Bhojpuri region were highly trained and skilled. That is why before the arrival of the East India Company in late 1801 it had become a big market for labourers skilled in sugar cultivation.

Mechanisation and change of routines make plantation workers' existing knowledge obsolete, also known as 'deskilling' (Braverman 1974). In economic development, deskilling is an obvious idea, but the speed with which knowledge disappears in industrial production is considerably greater than in plantation agriculture. The main reason for this is that British Indian plantation workers often still had the opportunity to grow their own living, even if it was only some rice, vegetables, pepper and other spices. So these workers did not completely disappear from their original economic environment like the peasants and artisans who were transformed into industrial workers. The relatively short contract period of 5 years (in Suriname) will also have limited the deskilling of agrarian human capital.

2.4 Recruitment

Who were recruited? The answer to this question depends on another answer, namely, how were they recruited? Underlying this question is the reasoning that the method of recruitment also says something about who was approached and thus recruited. The 'who' question also depends on where they were recruited and what happened to them afterwards.

Surinamese authors on recruitment are divided into two camps, both of whom, incidentally, are happy to refer to colonial sources. One camp says that recruitment was full of deception and cheating, while the other judges that such practices did occur but were not characteristic of the system (see, for example, Choenni 2016 for the polemic in this area). However, I am more interested in whether the workers recruited were linked to agriculture and how they were recruited. For it is through recruitment that they were transformed into workers.

In a detailed study, Alexander et al. (2018) analysed the various migration flows in British India from the second half of the nineteenth century onwards. Bengal, which included the recruitment areas of Surinamese indentured labourers, was then part of the British colonial empire along with Bangladesh and Burma. National borders did not exist and migration was a permanent feature throughout the area. This mobility had to do not only with British economic policies that allowed the transport of goods and people (via rivers, roads and railways), but also with the expansion of the colonial state, which needed more and more soldiers and civil servants. An important category in this 'domestic' mobility was formed by newly married women who (at a relatively young age) permanently exchanged their parental home for that of their spouse: '[B]y 1921, almost one in ten of India's population, or some 30 million people, were internal migrants ... more than the total number of journeys overseas migrants made *over the course of an entire century*' (Alexander et al. 2018, p. 35, italics in original).

The Netherlands and France could not directly draw on the British Indian labour reservoir. Experience in Suriname's neighbouring country, then known as British Guiana, where the immigration of these workers began as early as 1838, had yielded positive results. This led the planters in Suriname, through the Dutch government, to request that workers also be recruited in British India. At first, the British government refused because it considered the health conditions in Suriname poor (Hoefte 1998). In 1870, the Dutch and British governments, which were embroiled in a protracted military conflict over the control of Aceh (Sumatra), started negotiations. It is still unclear whether the Dutch government obtained permission to recruit workers by exchanging the Gold Coast (Ghana) for the British claim to Sumatra, but the two countries concluded an agreement in 1872 (Hoefte 1998, pp. 30–31).

The recruiting organisation consisted of several parties, who in turn employed others. Their recruitment activities were overseen by government agencies. I borrow from De Klerk (1953, p. 16) the following overview of parties involved in the organisation:

- Protector of Emigrants, in every embarkation port of India, appointed by the Indian government.
- Emigration agent, head of the emigrant depot of a particular colony, at an embarkation port of India appointed by the importing colony and approved by the British government.
- Subagent, head of a subdepot in the interior of India, appointed by the emigration agent.
- Recruiters, employed in India by the subagent(s) and authorised by the Protector of Emigrants.
- Agent-General, head of the Immigration Department in the immigrant colony, appointed by its own Colonial Administration or by the Supreme Administration in the mother country.
- Subagent (-general), in importing colony, assistant and deputy to the agent-general.

To these parties, De Klerk adds that in Suriname, the British consul also operated as a Protector of Immigrants. Some authors (Bhagwanbali 2010, pp. 46–56; Choenni 2016, p. 145ff) also list the remuneration that the various parties received, but this is a meaningless exercise because the recruitment costs and purchasing power are unknown.

Planters in Suriname directed their request for a specific number of workers to the agent-general, who passed the demand on to the emigration agent in British India. In practice, this was the Dutch consul in Calcutta. The primary task of the emigration agent was to ensure that recruitment and selection proceeded in accordance with the prevailing rules. The request for workers was passed on to the emigration agent, who employed subagents. These subagents employed a team consisting of clerks, doctors, guards and cooks. It was these subagents who had employed the actual recruiters, called ‘arkatiyas’. For each migrant supplied, the emigration agents and subagents received a premium, which incidentally varied with the price of food, the sex of the emigrant and the difficulty of recruiting them, e.g. due to calamities (Bhagwanbali 2010, pp. 52–53).

Bhagwanbali (2010, p. 47) reports that the subagents had established their subdepots mainly in the overpopulated cities. According to the guidelines, the subagent had to supply men between the ages of 10 and 35, women between 10 and 30, and children between 2 and 10. For every 100 men, the number of women had to be 40, while the number of children should not exceed 10%. Certain castes such as

Brahmins and Kshatriyas were considered unsuitable for plantation labour. Fakirs, beggars, people in poor health, the elderly and the handicapped were also barred.

Although the laws surrounding both emigration and immigration and the rights of indentured labourers were regularly amended, there was a latent and sometimes overt resistance to recruitment in British India. This resistance came from the population, local (village) police, British colonial officials, judges and Indian parliamentarians involved in the development and implementation of legislation (Fokken 2018). Officials were critical of recruitment practices and judges were not always satisfied with simple answers from the migrants who had to make procedural statements.

Initially, potential migrants were housed in subdistricts. Bhagwanbali (2010, p. 50) describes these as 'structure(s) of thick mud walls with a floor of rammed earth fitted with a roof of thatch or tiles'. Most subdepots had a common sleeping area for men and women. Sleeping benches were missing, as were separate toilets and bathing areas for men and women. In these subdepots, the men underwent a medical examination, which, incidentally, was not rigorous and mainly served to determine whether the person was suitable for fieldwork and long sea voyages, and to check for diseases.

The provisions of the employment contract were regularly amended over the years. The main provisions were that workers had the right to free transport and return, as well as free medical treatment, and that men received a daily wage of 60 ct. and women and children 40 ct. The contract further stipulated that the workers had to work 6 days a week, 7 h a day in the field or 9 h in the factory (Bhagwanbali 1996, Annexes 1, 5 and 17 for a summary of the provisions). Their freedoms were limited as workers were not allowed to leave the plantation without a pass. A contentious issue among colonial rulers was that planters and later district commissioners had the power to issue punishments. These powers were later granted to the judiciary. Nevertheless, for a long time the labourers were punished more severely (with fines, corporal punishment and imprisonment, with or without forced labour) and the planter only with a fine (De Klerk 1953, p. 124). After the expiry of the contract, workers could return to British India, renew the contract or (especially after 1895) settle as peasants.

As in the subdepots, dropouts occurred at the main depot in Calcutta. The extent of these dropouts is difficult to determine because figures for the period 1873–1916 are not complete and/or because the figures are total numbers of migrants assigned to different countries. But the available studies indicate a dropout rate in the main depot of over a fifth of the migrants recruited. The main reasons for the dropouts were that the recruited migrants were not found to be suitable for fieldwork (Emmer 1984, pp. 150, 155; cf. Bhagwanbali 1996, p. 121ff). Other reasons included illness and death, long waiting times (1–2 months), a change of intention (so-called 'deserters') and people who were 'claimed' by family.

The stay in the main depot is significant in three respects. The first was the signs of the migrants' suffering. They were, as it were, absorbed and confined—'a veritable human warehouse', according to De Klerk (1953, p. 67)—in procedures and explanations, and isolated from the outside world. Compensating for the 'detention' were free food, clothing and medical facilities. Thus, the stay in the main depot was not entirely 'bad'. Secondly, and more importantly for an understanding of later ethnic community formation in Suriname, the depot brought together people of different origins who entered into their first social relationships here. These would later take on the character of family relationships, known as *dipuá* and *jaháji* relationships (Gillion 1962). The third issue concerns the autonomy of migrants, which is best illustrated by a complaint by a British official that some migrants took advantage of the facilities offered by emigration agencies and then refrained from travelling overseas (Kumar 2017, p. 62). If there was any deception, it was among both recruiters and migrants.

An important effect of social interaction in the depots was what Lal (2006, p. 371) called 'the great leveler'. Unlike Seigel (2017), whose book, in which he explores the role of violence in reducing social inequality, has an eponymous title, Lal meant that the extant social divisions in the caste hierarchy were broken down. At registration in British India, people from low castes could pretend to be members of higher castes, Muslims could call themselves Hindus and vice versa, and people fleeing justice and family could adopt a different name, caste and origin. In other cases, names were changed to make detection by family or judiciary difficult. The changes in personalia and outward appearance were legion (Choenni 2016; Fokken 2018). Because of the anecdotal nature of these observations, it is unclear on what scale this transformation of personal identity occurred, although one author suggested it was on a considerable scale (Jayawardena 1963, p. 19). However, Bisnauth (2000, p. 47) pointed out that there was also a downward mobility, especially of members belonging to higher castes. He characterised this downward mobility process as *chamarisation*, the very opposite of upward mobility.

The first major cultural transformation, parallel to the change from peasant to labourer, thus began in the subdepots where migrants had the opportunity to change status and identity. It was part of the proletarianisation process.

Several authors (Bhagwanbali 1996; Choenni 2016; De Klerk 1953) have elaborated on the caste origin of the immigrants. The significance of this is unclear since the caste system in most of the countries that received indentured labourers did not last, i.e. it had no lasting effect on the social structure of British Indian communities. With regard to the jobs the labourers were performing, Bates (2017, p. 18) pointed out that 'someone might work as a sepoy, then as a domestic servant, then perhaps they could become a lascar or an indentured labourer, progressing finally to farming or assisting in or keeping a shop'. So any relation between castes and the emigrants' occupation became blurred. Although the Surinamese authors point out that there is no one-to-one relationship between caste and occupation, they do make this connection.

This relationship is discussed most extensively by De Klerk (1953, pp. 98–112). Based on a sample of four ships, transporting a total of 2637 people, he arrives at the following compressed classification: high castes (18%), middle castes (33%), low castes (9%), pariahs (22%), Muslims (18%) and unknown (0.5%). De Klerk (1953, p. 109) notes that the classification allows for some slack, but he is aware of the agrarian background of the labourers. Further on in his book (De Klerk 1953,

p. 111), he lists the declared occupations of adult passengers of two ships, 362 and 425 in number. His tables show that the number of peasants and field workers was less than half of the total number of passengers. That is most likely the outcome of the rapid change of jobs as Bates pointed out. As all these occupational groups were embedded in an agricultural environment, the practitioners arguably possessed some degree of agricultural human capital.

Lal (1983a, p. 72) also observes a discrepancy between caste and occupation in the United Provinces on the basis of the 1911 census and notes that a large proportion of traditional castes had left their original occupation. 'Thus less than 10% of Ahirs, Brahmans, Chamars, Kewats and Pasis practised their traditional occupation. The rest had taken to other occupations, principally agriculture. Over 70% of Ahirs, Brahmans and Kewats derived their livelihood from cultivating land' (Lal 1983a, pp. 71–74). This observation reflects the major shifts within the caste system and its diminished significance in indicating occupation. Irrespective of their caste, these people possessed agrarian human capital because they (partly) lived in, or were connected to, the agricultural sector.

The literature has frequently noted that peasants have numerous subsidiary activities that provide regular income (Goodman and Redclift 1981; Mintz 1973; Shanin 1984). A barber, shopkeeper or rice mill owner in a village, because of their embedding in the agrarian milieu, even when practising the profession full-time, usually also engages in farming, albeit for their own consumption. This is possible because they stay in one place and, unlike the transporter or vendor, are immobile. But even for those who were mobile, there was room after formal working hours and at weekends to do something on the plot alone or with the family. Moreover, the activities of such families were often intertwined with agricultural activities of relatives. Agricultural human capital was thus the property of a larger category of people than can be determined on the basis of caste or occupation.

2.5 The Crossing

The sea voyage was the second stage in the transformation from peasant to labourer. The voyage was significant because two developments occurred in this journey. The first was the change of social identity. The second was the creation of new relationships, known as *jahájis* (Gillion 1962). Both types of changes would later be important for ethnic group formation in Suriname.

Just before departure, passengers were asked (for the third time) whether they wanted to leave. Those who answered in the negative or who had doubts were turned away (Majumder 2010, p. 26). These people may have been misled, but that raises the question of why they had not made this known at the two earlier opportunities they had to make objections. It may also be that they had initially answered in the positive as they had been 'imprinted' and that at the time of departure, feelings of doubt prevailed. Neither can it be ruled out that they were simply ambivalent from the start and changed their minds at the last moment. Such ambivalence will have played a role for several labourers—including those who later decided to stay in Suriname—without affecting the decision at the moment of departure.

The crossing took place in sailing and steam ships. The journey by sailing ship took 3 months and by steam ship 6–8 weeks. During the first 5 years, both sailing and steam ships were used, then until 1907 predominantly sailing ships and from 1907 only steam ships (De Klerk 1953, p. 73). The voyage started in Calcutta and had a break at St Helena to stock up on food. Sometimes other ports were called at to avoid storms, to take in drinking water and food, or because of calamities on board such as impending rebellion (Photo 2.1).

The arrangement of ships is receiving increasing attention in the literature. Kumar (2017) points out that in the period 1836–1866, transport switched from sailing to steam ships. He speaks of a 'revolution in sea transport', which he describes as follows: 'The sea passage became significantly more comfortable in the nineteenth century ... Nineteenth-century ships were not only spacious, well-ventilated and carrying more passengers, but also faster ...' (Kumar 2017, p. 80). In contrast, Fokken (2018, pp. 105–116) highlights how the design of the ships took little account of the migrants' origins and affected life on board. The ships had an upper deck, a section in the middle and a third, lower deck. The migrants stayed in the middle section, which was divided into three compartments: for single men, couples

Photo 2.1 British Indian indentured labourers on the deck of a ship at the Suriname river Paramaribo returning to British India

(with or without children) and single women. Passengers were expected to spend their time above deck during the day.

While some authors argue that caste and other restrictions disappeared during the voyage (e.g. De Klerk 1953), Fokken (2018, pp. 110–111) points out that new hierarchies emerged during the voyage. From the passengers, ship personnel were chosen, including a cleaner, cook, barber, tailor and supervisor, who, incidentally, received remuneration for their work. Fokken deduces that some migrants enjoyed greater freedom of movement and held more power than others. This was probably the case during the voyage, but it is by no means certain that this new hierarchy was perpetuated in Suriname. In all likelihood, it was a temporary social arrangement that lost its significance at the end of the voyage.

In addition to the social changes during the voyage, demographic changes also took place. De Klerk reports mortality on board, with seasickness and epidemics being the main causes. The two highest numbers of deaths per ship were 42 and 48. He lists 15 ships with above-average deaths that ranged in number from 11 to 36. 'Even without certain epidemies being involved, mortality was high on most emigrant ships. A figure of 5-10 was quite normal' (De Klerk 1953, p. 76). Favourable outliers were three ships with no deaths, three ships with one death and one ship with two deaths. De Klerk also mentions 12 ships that had 103 births. He explained the births as resulting from pregnancies from relationships that violated caste rules and other restrictions prevalent in British India. These couples or single pregnant women were on the run and would sometimes have given birth prematurely during the voyage due to the turbulent weather conditions.

Little is known about social life on the ship, except that it was limited. It was a monotonous voyage and many could not cope with the loneliness and other forms of deprivation (De Klerk 1953). Food and health were regularly checked by the accompanying ship's doctor, but he could not prevent sickness and sometimes epidemics. Suicides also occurred. Social life was tightly disciplined and routine and mainly took place above deck. Mutual amusement was encouraged. But this intercourse, according to Fokken (2018, p. 116), was limited because '[t]he physical layout of the ship and the rules and routines applied by the surgeon superintendent determined what the migrants' day looked like and whom they were able to socialise with'. Earlier in her book, she called the limited space for movement 'moving within confines' (Fokken 2018, p. 109). During the sea voyage, the first development, the change of personal identity, was completed.

The second development, the creation of new relationships, has also received attention in the literature, but its historical and anthropological significance has hardly been recognised. It concerns the bonds between the migrants that were already formed during the stay in the main depot in Calcutta and later during the sea voyage. The migrants themselves used the term *jaháji bhái* (literally: brother of the journey), which elevated the travelling companion to brotherhood. Later, feminist authors suggested the term *jahájin* (sister of the travelling companion) (Mohan 2007). *Jahájis* were support and companions, not only in performing work, but also for social life, especially to combat loneliness, organise companionship, perform rituals and ensure protection from other people, which women especially needed.

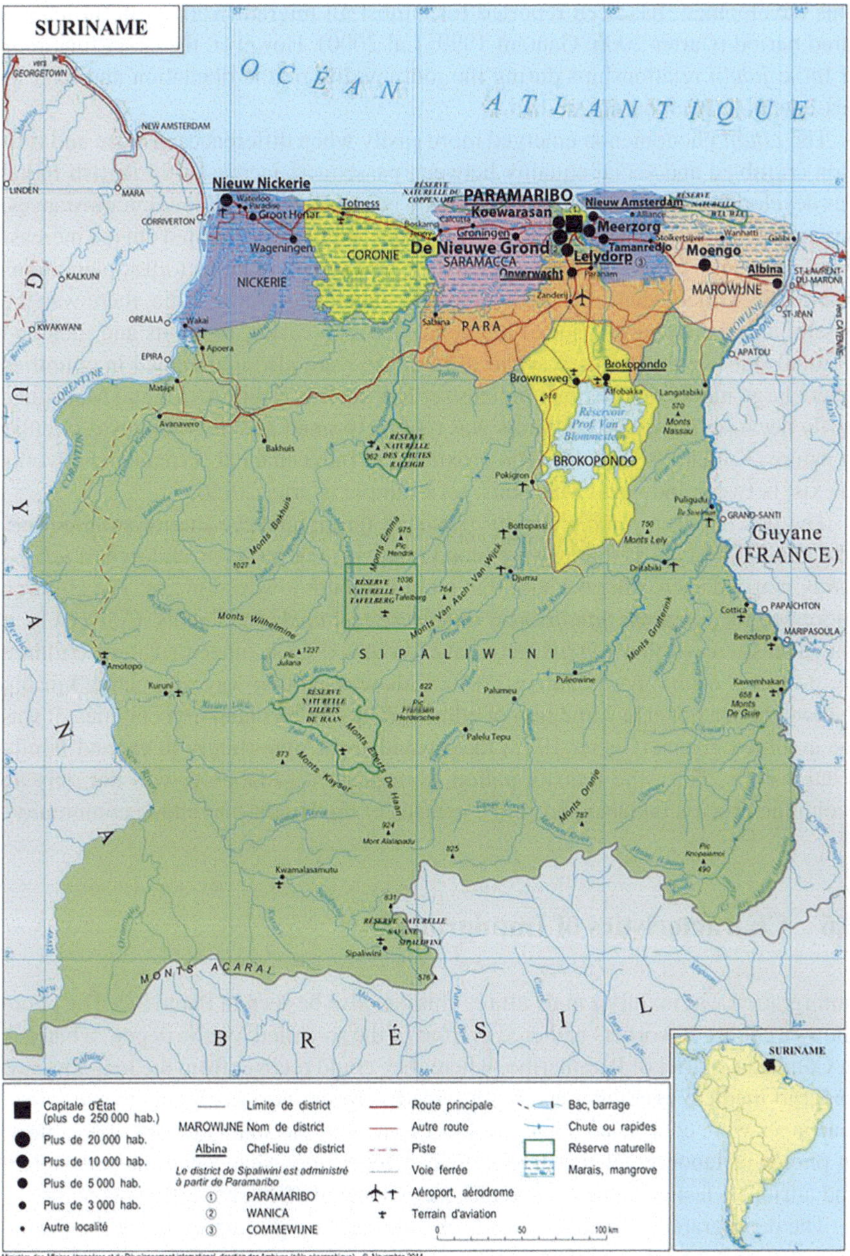

Ministère des Affaires étrangères et du Développement international, direction des Archives (pôle géographique) © Novembre 2014

Map 2.1 Map of Suriname

This development has been reported for almost all migrant workers in the indentured period (Carter 2006; Gautam 1999; Lal 2000). However, the exact functions of these *jaháji* relationships during the journey, life on the plantation and later as peasants have been hardly studied.

The *jaháji* phenomenon emerged more easily when differences in caste and religion crumbled and social equality between passengers increased. In British India, this development had already set in when the peasants, due to various circumstances, became labourers. Remember the study by Lal (1983a), who pointed out the increasing discrepancy between occupation and caste. In work—the factories, plantations and public works—there was little scope for caste rules. Yet in India, there was the possibility of maintaining social relations (between castes, religions and men and women) outside work. These social differences were also maintained in countries where large numbers of British Indians settled, such as South Africa and Mauritius, partly because the distance to India was relatively small and migration was circular in nature (Kumar 2017, p. 43). The proximity to India enabled permanent networks to exist between India and Mauritius, for example (Carter 2009).

The ties between *jahájis* usually persisted throughout the immigration period. The *jahájis* behaved as if they were blood relatives. For instance, children of *jahájis* rarely married each other because that would amount to social incest. The children addressed their parents' *jahájis* as if they were real uncles and aunts. Conversely, it was also the case that *jahájis* may have promised each other that their children would marry each other (Gautam 1999). This was another way of creating kinship that would promote the formation of ethnic groups in the plantation colonies. It was the assumed givens (Geertz 1971) that provided the opportunity to expand family relationships. Thus, the transformation from peasant to labourer was not only an economic process but also part of the genesis of the Hindustani ethnic community.

2.6 Characteristics of Immigrants

Immigration was mainly a male affair. This can also be seen in Photo 2.2. The photo shows a cohort of workers gathered in front of the so-called 'coolie depot', a barrack in Combé, a district of Paramaribo, where they could recover from the long journey, undergo medical examinations, be provided with the necessary papers and where marriages were contracted. Note the sober attire and the small amount of jewellery. In photos of labourers departing for British India they wear much more jewellery and attributes testifying to their wealth (see especially Choenni 2016).

The demographic characteristics of the immigrants are important for the formation of formal or informal families that will later become the units of small-scale agriculture. Chart 2.1 shows the spread of arrivals during the course of immigration over the period 1873–1916, i.e. a period of 43 years.

The immigrants came to Suriname in 64 ship transports, the majority as individuals and not as families. The figures in the graph (see Appendix 1) incidentally add up to 34,286 people, while De Klerk mentions the figure of 34,304. The graph

Photo 2.2 Arriving indentured labourers gathered at the coolie depot in Paramaribo, ca. 1885

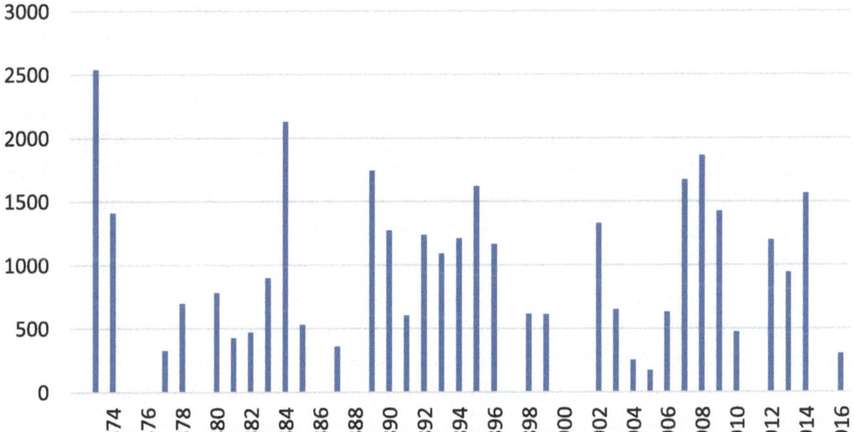

Chart 2.1 Migrants by numbers and year of arrival, 1873–1916. (Source: Adapted from De Klerk 1953, pp. 71–73)

shows large differences that are best illustrated by the lowest and highest number of passengers. On average, 535 people arrived per ship. The highest number of immigrants arrived in the first year, 1873. In the following year, there were just over half the number of arrivals. With the arrival of the 25th ship on 23 November 1895, more than half of the immigrants had arrived in Suriname. In other words, in the first half of the immigration period the supply was slightly larger than in the second half.

The graph further shows that there were periods when there was no supply of new workers. The longest period of stagnation in supply was between 1874 and 1877 (39.5 months). Other periods of stagnation were between April 1899 and January 1902 (a period of 33 months), from 27 January 1910 to 7 April 1912 (27 months), from 14 January 1885 to 18 January 1887 (24 months), from 18 January 1887 to 3 January 1889 (also 24 months), and from 10 January 1903 to 14 December 1904 and 4 June 1914 to 24 May 1916 (23 months each). Also striking is that, apart from the years 1873 and 1884, it cannot be said that the numbers of immigrants decreased, nor that the intervening periods increased.

The longest period of stagnation was due to the suspension of immigration, because in the first year and a half, 797 of the 3833 arrivals had died, and 30–40% of the immigrants on the plantations were in hospitals. De Klerk (1953, pp. 119–20) attributes the high mortality and morbidity to a confluence of very different factors, the main cause of which was poor selection during recruitment in British India. In Bihar and eastern Uttar Pradesh, the famine of 1873–1875 would have weakened the people and, combined with hasty recruitment, the selection was unfortunate. In addition, De Klerk mentions heavy plantation labour, doctors' unfamiliarity with specific needs, climate change, inadequate housing and poor care for the sick. This explanation is unconvincing because immigration in British Guiana and Trinidad, which had started several decades earlier, had also been suspended after the first arrivals due to poor health conditions (Laurence 1994, p. 3).

As mentioned before, women were under-represented among the labourers. Bhagwanbali (1996, p. 96) provides a gender breakdown of the immigrants who arrived in Suriname. His figures of the total number of immigrants that arrived differ considerably from those of De Klerk (1953, compare Annex 2.1 with Annex 2.2). Bhagwanbali's overview, which covers the period 1881–1916 during which 29 ships arrived in Suriname, shows that the proportion of women per ship ranged from 21% to 31% and the overall average was 27% (Annex 2.2). One-ninth of the migrants fell into the 0–10 age group, one-seventh into the 10–20 age group, almost two-thirds into the 20–30 age group and the rest were over 30 years old (Bhagwanbali 1996, p. 135). A demographic equilibrium might have been restored if women had been significantly involved with partners from outside the group, or if men had been significantly more likely to die or return to British India.

The latter was difficult to ascertain. After the period from 1878 (5 years after the first batch arrived) to 1926, 11,512 migrants were transported back to India, about a third of the total number of migrants that had landed. The figures are indicative because it is not clear how many of them left Calcutta, how many arrived alive, how many died during their stay in Suriname, how many were born during the journey and in Suriname and left for British India, and how many left via British Guiana or other routes. De Klerk (1953, p. 159) mentions that in 1948, 1500 were still entitled to free return passage. This means that these workers renewed their contracts often or worked as peasants, and, unlike in Jamaica and Fiji, for example, that the validity of their rights was not limited.

The question is how the demographic balance recovered. The data on returns do not provide an exclusive answer on this. Choenni (2016, p. 509) lists the gender of adult returnees (see Annex 2.3). His figures differ from those of Bhagwanbali (1996, p. 96) both in terms of numbers and year of departure and are therefore difficult to compare. Choenni gives three categories of returnees (men, women and children), while Bhagwanbali provides only the gender of the arriving workers. Moreover, Choenni's period covers 1878–1921, while Bhagwanbali's table spans 1881–1916.

Over the entire immigration period, the proportion of women arriving was 27% and that of women leaving was 20%. The difference between arriving and departing women was just 7 percentage points. That the demographic balance had recovered based on the departure of slightly fewer women is unlikely. Rather, the large discrepancy between the numbers of men and women makes it plausible that men had lived much longer without a partner or had recruited wives from the adjacent British Guiana, which harboured a larger British Indian population.

2.7 Conclusions

This chapter has argued that the inhabitants of north-eastern British India had a long tradition of migration. A large part of the population emigrated and often returned. The mobility of the population was the result of British capitalism that marginalised peasants and destroyed crafts and other professions. In addition to this disruptive influence, prolonged droughts played a role. These compelling circumstances fostered circular migration and a migration culture in recruitment areas. The existence of this culture contradicts the assertion that the workers were ignorant of the places of destination, of the work they had to do and that they were even gullible. It may be true that migrants sometimes had a vague idea of their destination country, that they had exaggerated expectations and were often disappointed. But exaggerated expectations are intrinsic to emigration, be it the American Dream or migration to European welfare states.

The indentured labourers were mainly peasants, agricultural labourers and artisans who had changed jobs several times. They possessed knowledge of agriculture, were embedded in the agrarian environment and were familiar with crop cultivation. They thus possessed agrarian human capital. This feature of proletarianisation has received little attention in the literature, in all likelihood because the implicit assumption was that the workers were recruited for the sake of their muscle power, i.e. to replace slaves. The Caribbean literature in particular has concentrated on the selection of labourers where suitability for field labour was the main criterion.

The perception that recruits were misled on a massive scale is contradicted by three key pointers that indicate the prevailing migration culture, namely the mobility of the population, knowledge of recruitment practices and related procedures, and knowledge of migration destinations. The existence of a migration culture suggests that the recruitment and migration of labour was not an individual but a collective event that affected the whole area. The system had some controls to prevent deception on a large scale. The view that deception occurred en masse cannot be reconciled with the use the recruits made of the facilities of the emigration agencies and denies the migrants a will of their own.

The recruitment and sea journey were not just an economic process. They also included a change in personal and cultural identity and engagement in new relations. These social changes could only take place when traditional social barriers (between castes, religions, and men and women) crumbled. This took different forms: an upward social mobility for the lower castes and other disadvantaged, a downward mobility of members of higher castes and an individual change of recorded religious, family and regional features. This equalisation of social statuses began as early as registration and continued during the stay in the depots and during the crossing. The proletarianisation of the peasants was thus not only an economic process and social schism but also represented the onset of new ethnic community formation.

References

Alexander C, Chatterji J, Jalais A (2018) The Bengal diaspora. Rethinking Muslim migration. Routledge, London

Bates C (2017) Some thoughts on the representation and misrepresentation of the colonial south Asian labour diaspora. South Asian Stud 33(1):7–22. https://doi.org/10.1080/02666030.2017.1300372

Bhagwanbali R (1996) Contracten voor Suriname. Arbeidsmigratie vanuit Brits-Indië onder het indentured-labourstelsel, 1873–1916. Doctoral dissertation, Radboud University (Nijmegen)

Bhagwanbali R (2010) Hindoestaanse migranten onder het indentured labour systeem naar Suriname, 1873–1916. De nieuwe awatar van slavernij. Amrit, Den Haag

Bisnauth D (2000) The settlement of Indians in Guyana 1890–1930. Peepal Tree Press, Leeds

Braverman H (1974) Labor and monopoly capital: The degradation of work in the twentieth century. Monthly Review Press, New York

Carney J (2001) Black rice. The African origins of rice cultivation in Africa. Harvard University Press, Cambridge/London

Carter M (1995) Servants, sirdars and settlers. Indians in Mauritius 1834–1874. Oxford University Press, Delhi

Carter M (2006) Mauritius. In: Lal B, Reeves P, Rai R (eds) The Encyclopedia of the Indian Diaspora. Singapore, Editions Didier Millet, pp 262–272

Carter M (2009) Subaltern networks in a colonial diaspora: A study of Indian migrants and Mauritius. In: Rai R, Reeves P (eds) The South Asian Diaspora. Transnational networks and changing identities. Routledge, London, pp 45–56

Choenni C (2016) Hindostaanse contractarbeiders 1873–1920. Van India naar de plantages in Suriname. LM Publishers, Volendam

De Haan A (2002) Migration and livelihoods in historical perspective: A case study of Bihar. India Journal of Development Studies 38(5):115–142

De Klerk CJM (1953) De Immigratie der Hindostanen. Urbi et Orbi, Amsterdam

Emmer P (1984) Een goede keus? De migratie van vrouwelijke contractarbeiders van Brits-Indië naar Suriname. OSO, Tijdschrift voor Surinaamse Taalkunde, Letterkunde en Geschiedenis 3(2):213–234

Emmer P (1985) De werving van kontraktarbeiders in India voor tewerkstelling in Suriname, 1872–1916. OSO, Tijdschrift voor Surinaamse Taalkunde, Letterkunde en Geschiedenis 4(2):147–157

Emmer P (1986) The meek Hindu: The recruitment of indentured labourers for service overseas 1870–1916. In: Emmer P (ed) Colonialism and Migration: Indentured Labourers Before and After Slavery. Martinus Nijhoff, Dordrecht, pp 187–207

Fenton S (2003) Ethnicity. Key concepts. Polity Press, Malden

Fokken M (2018) Beyond being koelies and kantráki. Constructing Hindostani identities in Suriname in the era of indenture 1873–1921. Doctoral dissertation, University of Groningen, Groningen

Gautam M (1999) The construction of the Indian image in Surinam: Deconstructing colonial derogatory notions and reconstructing Indian identity. In: Mahin G, Narine D (eds) Sojourners to Settlers: Indian Migrants in the Caribbean and the Americas. Windsor Press, New York, pp 125–181

Geertz C (1971) The integrative revolution: Primordial sentiments and civil politics in the new states. In: Welch C (ed) Political modernization: A reader in comparative political change Belmont. Duxbury Press, CA, pp 198–218

Gillion KL (1962) Fiji's Indian Migrants: A History to the End of Indenture in 1920. Oxford University Press, Melbourne

Goodman D, Redclift M (1981) From peasant to proletarians: Capitalist development and agrarian transitions. Basil Blackwell, Oxford

Grierson GA (1883) Report on colonial emigration from the Bengal presidency. Government of Bengal, Bengal

Hassankhan M (2013) Kahe Gaile Bides – Why did you go overseas? An introduction in emotional aspects of migration history: A diaspora perspective. In: Mahabir K (ed) Caribbean issues in Indian diaspora. Serial Publications, New Delhi, pp 3–35

Hira S (2000) Terug naar Uttar Pradesh. Op zoek naar de wortels van Surinaamse Hindostanen. Amrit, Den Haag

Hoefte R (1998) In place of slavery. A social history of British Indian and Javanese laborers in Suriname. University Press of Florida, Gainesville

Jayawardena C (1963) Conflict and solidarity in a Guyanese plantation. The Athlone Press, London

Kumar A (2017) Coolies of the Empire. Indentured Indians in the sugar colonies 1830–1920. Cambridge University Press, Cambridge

Lal B (1983a) Girmitiyas. The origins of the Fiji Indians. Australian National University Press, Canberra

Lal B (1983b) Indian indentured historiography: A note on problems, sources and methods. Pacific Studies 6(2):33–50

Lal B (2000) Chalo Jahaji. On a journey through indenture in Fiji. Australian National University Press, Canberra

Lal B (2006) Fiji. In: Lal B, Reeves P, Rai R (eds) The Encyclopedia of the Indian diaspora. Singapore, Editions Didier Millet, pp 370–382

Laurence K (1994) A question of labour: Indentured immigration into Trinidad and British Guiana 1875–1917. Ian Randle Publishers, Kingston

Majumder M (ed.) (2010) Kahe Gaile Bides (Why did you go overseas?). In: On Bhojpuri migration since the 1879's and contemporary culture in Uttar Pradesh and Bihar, Suriname and the Netherlands. Mango

Mintz S (1973) A note on the definition of peasantries. J Peasant Stud 1(1):91–106

Mohan P (2007) Jahajin. HarperCollins, Delhi

Narayan B (2017) Culture and the emotional economy of migration. Routledge, London

Northrup D (1995) Indentured labor in the age of Imperialism, 1834–1922. Cambridge University Press, Cambridge

Price R (2007) Some anthropological musings on creolization. Journal of Pidgin and Creole Languages 22(1):17–36

Roopnarine L (2003) East-Indian indentured emigration to the Caribbean: Beyond the push and pull model. Caribb Stud 31(2):97–134

Roy T (2018) Inequality in colonial India. Economic History Working Papers, no. 286. The London School of Economic and Political Science

Seigel W (2017) The great leveler. violence and the history of inequality from the stone age to the twenty-first century. Princeton University Press, Princeton

Shanin T (ed) (1984) Peasants and peasant societies. Penguin, Harmondsworth

Tinker H (1974) A new system of slavery: The export of Indian labour overseas 1830–1920. Oxford University Press, London/New York

Yang A (1979) Peasants on the move: a study of internal migration in India. J Interdiscip Hist 10(1):37–58

Chapter 3
Labourers on the Plantations

Abstract This chapter deals with the experiences of the labourers on the plantations. It has been overlooked that the labourers' experience on the plantations varied, but for the majority of them, it was a push factor. Hence, the Girmitiyas settled en mass as peasants when conditions were favourable. The argument is structured as follows. Section two outlines the centrality of the tasks on the plantations, their allocation and the restrictions imposed on labourers. The tasks had direct effects on health and family formation, which is the subject of Sect. 3.3. This triptych of tasks, health and family formation determined, given the plantation regime, the labourers' earning capacity. Actual earnings are also a major bone of contention and are discussed in Sect. 3.4. Where one author speaks of scandalous exploitation, another talks about large amounts of savings. Since the issue of savings is part of the controversy over wages, that issue is addressed in Sect. 3.5. Finally, in Sect. 3.6, the cultivation of crops for personal consumption and sale, i.e. the entrepreneurship of plantation labourers is discussed.

Keywords Proletarianisation · Family formation · Earning capacity · Wages · Savings · Entrepreneurship

3.1 Introduction

The proletarianisation of labourers was completed when British Indians became plantation labourers. Because by far the majority of historical research has been descriptive, hardly any links have been made between labour life on the plantations and subsequent developments. This is a result of the lack of a theoretical perspective, so that most of the discussions have the character of a directionless enumeration of facts. At the same time, this very lack of a theoretical lens gives the authors great freedom to use their personal vision as a framework. Examples include Bhagwanbali (2010), who draws attention mainly to the darker side of the history of plantation labour, and Choenni (2016), who, along with De Klerk (1953) and Emmer (1984, 1985a, b), views indenture as a blessing.

R. Gowricharn, *The Girmitiya Peasants in Suriname*,
https://doi.org/10.1007/978-3-031-67961-2_3

What were the labourers' experiences on the plantations? Were the experiences in the plantation colonies really a new system of slavery? Experiences will no doubt differ from colony to colony, but given the major similarities as discussed in the introduction to this book, those differences cannot have been great. The similarities suggest that, by and large, the same forces were at work on the plantations. Put differently, did experiences on the plantations contribute to later establishment as peasants? That is the core question of this chapter.

The key question revolves mainly around working life. Characteristic of this dimension of life were the tasks that determined the severity of work and varied with the product grown. Since the vast majority of historical descriptions (including those in the English-language literature) are based on sugar plantations, the differences from cocoa, coffee, timber, cotton, tobacco and other plantations hardly come into focus. Other issues arising from work on the plantations were earnings, health and savings. There is no agreement on any of these issues, or to be more precise: the historical judgment tilts towards the positive for most authors (Choenni 2016; De Klerk 1953; Emmer 1984, 1985a, b), while other authors are less outspoken (Fokken 2018; Hoefte 1998) and a few are radically negative (Bhagwanbali 2010). By connecting these issues, I try to answer the question of which factors facilitated settlement as a peasant. The reason for raising the labourers' experience is thus not to judge the system nor to denounce it. My point is that the labourers' appreciation, i.e. their 'moral experience', is crucial in explaining their preference to settle as peasants.

This chapter addresses the labourers' experiences on the plantations. Those experiences are gauged from work ability, health, wages, savings and the ability to sustain agricultural human capital. Wages have been presented in the literature either as being generous or as evidence of exploitation. This would explain the savings or lack thereof. This chapter makes it clear that the sources of income were not limited to wages, making it a misconception to relate savings exclusively to wages earned. The labourers were also active as agriculturists, so their proletarianisation was not complete. Income and human capital were not the same for all labourers, but logically and chronologically, these elements are important in understanding later settlement as peasants. In other words, labour experiences at the plantations and future prospects were closely linked.

In the next section, I discuss the labour process. Central here are the tasks, their allocation and the restrictions imposed on labourers. The tasks had direct effects on health and family formation, which is the subject of Sect. 3.3. This triptych of tasks, health and family formation determined, given the plantation regime, the labourers' earning capacity. Actual earnings are also a major bone of contention and are discussed in Sect. 3.4. Where one author speaks of scandalous exploitation, another talks about large amounts of savings. Since the issue of savings is part of the controversy over wages, I address that issue in Sect. 3.5. Finally, in Sect. 3.6, I consider the cultivation of crops for their own consumption and for sale, i.e. the entrepreneurship of plantation labourers. In the closing paragraph the findings are interpreted.

3.2 The Labour Process

The migrants had signed a labour contract in which the main provisions were that they committed themselves for 5 years to work 313 days a year, 6 days a week and 7 h a day in the field or 10 h (other sources say 9 h) in the factory. Should they become permanently unfit for work and this was not their fault, they would be sent back at the Dutch government's expense. The number of hours worked could be increased if the contractor had deserted or served a prison sentence, and reduced if they had worked overtime. The number of compulsory working days was reduced by the number of generally recognised holidays in the colony, and by 32 holidays for Hindus and 16 holidays for Muslims. On these days and when a worker was not assigned work, he did not need permission to leave the plantation (De Klerk, p. 131ff).

Many of these religious holidays have fallen into oblivion, but Choenni (2016, pp. 705–706) gives an overview of Hindu holidays. Most festivals took place within the confines of the plantations, but some attracted the attention of the authorities because they could get out of hand (Fokken 2018, p. 189ff).

According to Fokken (2018, p. 150), labourers were divided into classes: the first class included physically fit men aged 16 and above, the second class comprised boys aged between 10 and 15 and all women older than 10, and the third class consisted of boys and girls aged below 10. Each of these categories received a different wage. Fokken argues that this classification allowed wages to be kept low because a maximum wage was formally agreed upon and not a minimum wage. In contrast, Snellen (1933, p. 53), Bhagwanbali (2010, p. 152) and De Klerk (1953, p. 132) speak of an agreed minimum wage. As will be shown later, labourers could earn more than agreed in the contract. Thus, the wages mentioned must have constituted a lower limit. 'Where no work is provided on a working day, the minimum wage for a full day's work will nevertheless be paid. For task work, the wage shall be the same as that of labourers not under contract on the same or neighbouring plantations and not less than the minimum wage' (De Klerk 1953, p. 132). Women were given lighter work tasks and also earned less.

Most authors claim that the work on sugar plantations was heavier than that on other plantations, but comparisons with, for example, timber plantations or the shaded coffee and cocoa plantations are scarce. Comins (1892, p. 5) writes: 'Cacao estates are preferred by the coolies to sugar estates because (a) there is not the same discipline enforced; the labourers being in small numbers do not obtain or need the same organised control; (b) the pay is generally somewhat higher and is more capable of expansion by the labourer himself; (c) the work is not so hard, e.g. trimming trees and picking beans; (d) it is done mostly in the shade and not in the heat of the sun; (e) as the estate is not very large, the labourer is not so far from his own hut; (f) there are opportunities for growing roots, "ground provisions", and plantains among the cacao trees, which cannot be done in a "cane piece".'

Bhagwanbali (2010, p. 143) gives a brief description of the tasks: '… a work task on a cotton plantation consisted of weeding, planting cotton pieces, picking, cleaning and beating. Work tasks on cocoa and coffee plantations usually consisted of weeding, transplanting young cocoa and coffee plants, making holes for transplanting, picking and mauling cocoa and coffee. The work tasks on a sugar plantation were usually mining cane holes, planting and covering cane tops, weeding young cane, trassen [Dutch, removing leaves from the cane], cutting and carrying the cane; in the factory, one was usually employed in the sugar preparation process.' For sugar plantations, Hoefte (1998) points out that the thickness of the soil, the density of the vegetation and the weather conditions, which depended on the seasons, all had to be taken into account. The tasks that made the heaviest demands but enabled labourers to earn the most were '[c]hopping trees and brushwood ..., followed by such jobs as skilled factory works (panboiling, carpentry), railroad maintenance, digging, cutting cane, unskilled factory labour, manuring, forking, banking and weeding' (Hoefte 1998, p. 115).

The daily wages were based on the tasks, which could vary by type of plantation and also within one plantation. The tasks on sugar plantations were the most labour-intensive and therefore the heaviest. There was a lot of discussion among the colonial stakeholders, especially the governors and planters, but also among civil servants, about the gravity of the tasks. Most opinions inclined towards the view that the tasks were too heavy for the labourers and that they were determined unilaterally by the planter (see Hoefte 1998, Chaps. 7 and 8, for a discussion).

A 1909 report on the conditions of plantation labourers, which incidentally was based on the conditions of Javanese labourers, stated that the tasks were composed on the basis of a 10-h working day, a norm from slavery (Hoefte 1990). But according to Snellen (1933, p. 53), who worked in the Department of Agricultural Economic Affairs and wrote a doctoral thesis on the supply of labour, the load of tasks had been proportionally reduced to seven and 10 h for field and factory labour, respectively. Since most contract labourers throughout the Indian diaspora were employed on sugar plantations, experience gained on these plantations dominates the literature.

According to Bhagwanbali (2010, p. 141), the working day started at 4 am. The labourers prepared their meals and after eating they gathered their tools. Hoefte (1998, p. 114ff) states that labourers were divided into labour gangs, taking into account the type of work, seasons, gender and age. The working day formally ended 7 or 10 h later, but the extent to which this provision was observed is not clear. In practice, labourers could be forced to work longer, for instance by rejecting the tasks performed (see Bhagwanbali 2010 for many examples). Some authors suggest that such incidents occurred only on distressed plantations (De Cock Buning 1926; Snellen 1933). But making labourers work longer did not only stem from the economic situation of the plantation. Given the differences in power relations and the legal vulnerability of the worker, as well as the mentality of the plantation authorities, more factors probably led to arbitrariness and exploitation (Choenni 2016, Ch. VII; Hoefte 1998, pp. 114–124).

The plantation had a hierarchy that Hoefte (1998, p. 94ff) classified, on the basis of a study of the Mariënburg sugar plantation, into a group for fieldwork, a department for the factory and personnel for office work. From the free labourers, consisting mainly of Creoles, some of the heads of labour gangs, also known as *sardárs*, who were in touch with the field labourers were recruited through foremen. Indications that higher castes were more likely to be considered for these management positions are non-existent. Sometimes names of these foremen are mentioned that indicate a higher caste descent, but it also happened that members of lower castes were appointed to these positions. As mentioned before, names are not a reliable indicator as they underwent changes during registration in British India. It was mainly men who compiled the tasks and distributed and supervised the work.

The treatment of labourers was a matter of constant concern. This applied especially to the composition of heavy tasks. Most complaints filed were related to labour and wage issues (Bhagwanbali 2010, p. 168ff; Choenni 2016, p. 450ff; Fokken 2018, p. 155ff). These formal complaints represent the tip of the iceberg as labourers were not always given the required passes to leave the plantation to lodge complaints. The treatment was also reflected in daily dealings in the field. British consul Annesley (1880–1885) characterised the treatment by the *sardárs* as follows: 'They are always barking and biting at the coolies, continually making use of filthy expressions in Hindustani, which they learned by heart without hardly knowing their meaning' (quoted in Fokken 2018, p. 153). Such treatment by the plantation authorities was a structural feature, but in official annual reports the treatment was presented more favourably.

The sexual molestation of women was, remarkably, hardly discussed by an earlier generation of authors (De Klerk 1953; Snellen 1933). However, the subject has been raised by later authors, albeit not all emphasising it. Molestation by plantation authorities was common and was even the cause of many a violent incident and of labourers' uprisings on the plantations (Choenni 2016, p. 584ff; Hoefte 1998, Ch. X). In most cases, women have been portrayed as defenceless victims, but recent literature (including international literature) increasingly reports that women had agency, i.e. possessed some independence, and were not always victims. This agency would be reflected in the fact that some women had multiple husbands and were guided by financial gain (Choenni 2016, p. 638; Fokken 2018, p. 170ff; Kumar 2017, p. 176ff).

These authors' focus on women's sexual behaviour to illustrate agency is curious. Disguised prostitution and money-hungry women will undoubtedly have occurred, although the scale of this occurrence is unknown. The agency of women is only illustrated on the basis of sexual intercourse. That is peculiar as there were also men who had their wives perform sexual services as reported by a former official in Suriname (Wesenhagen 1886, pp. 49–50). Since the hub of life on the plantation revolved around labour, both for women and men, it is more appropriate to show agency in that area. It remains unclear what the meaning of sexual agency is and why it is brought in.

The labour process on the plantation was generally hard, but that varied with the product grown. As mentioned before, in the literature, the experience on the sugar plantations dominates. Labourers formally enjoyed a remarkably large number of days off, but whether, and if so, how these were granted in practice is unclear. The reason for this doubt is that the plantation authorities had the ability to evade numerous contract provisions and bully the labourers. In contrast, labourers had few opportunities to file complaints. The role of women constitutes a specific element in the treatment. They were often sexually molested, but it is also often reported that they themselves could sometimes take the initiative to seduce men.

3.3 Health and Family Formation

The issue of health is important for two reasons. First, it concerns a very concrete price paid by labourers. This is not to say that medical illnesses were entirely due to working conditions. Numerous health issues could be traced back to physical conditions and these would not have been different in British India, rural or urban. But it is true that health may have been aggravated or mitigated by specific conditions in Suriname, such as the poor quality of care provided, housing, nutrition and the work regime. Health has an influence on labour performance and earnings, so it cannot be absent from an assessment of labour. The second reason is that population growth (the balance of birth and death rates) is partly determined by the health of men and women. Restoring the sex ratio is important for achieving family formation, which will later determine the labour capacity of peasants.

What was the health status of the labourers? Already during the first shipload of 1873, it was apparent that it left much to be desired. High mortality and morbidity rates led to a suspension of immigration. The most common causes of death were ulcers, syphilis and anaemia. De Klerk (1953, p. 119ff) discusses a number of partial explanations but attributes the high mortality to the weakening of the population due to famines and the poor recruitment of labourers in British India. Especially in the decade after the first suspension, Dutch, British and Surinamese governments expressed their views on the issue. Unlike De Klerk, the British and Dutch governments believed that the causes of the high mortality had to be sought in the poor nutrition and medical care in Suriname (see Bhagwanbali 2010, pp. 122–127 for a discussion).

The health situation was not only a matter of great concern in the first decade of immigration. It also remained a hot topic in the decades that followed, as records and correspondence between district commissioners, medical inspectors and other officials show (Bhagwanbali 2010, p. 131ff). In 1915, at the end of immigration, British inspectors McNeill and Lal (1915, p. 42) wrote of the plantation labourers: 'The principal diseases are malaria, ankylostomiasis, pneumonia, and tuberculosis ... The average annual sickness from all causes ... has varied in recent years from 20 to 28 days per head ...' That the health of immigrants did not improve was mainly due to poor nutrition and hygiene, inadequate medical care and heavy work. These conditions were structural and not characteristic of any specific period or group. Fifteen years later, the government's chief medical officer would observe the same poor state of health (Lampe 1927).

Housing reflected the circumstances the labourers lived in and their influence on their physical condition. Formally, immigrants were entitled to a space three metres long, two metres wide and two and a half metres high in which no more than three people or one family could be housed. The dwelling had to be painted inside and provide sufficient light and air, while communal kitchens and latrines had to be adequate in size and discharge. The practice, however, was different. De Klerk (1953, p. 137) wondered how, despite the harsh reality and the wrecked material, the unhygienic and primitive practices of occupants, especially the lower castes, infestations of vermin and tropical rains combined, a physically and mentally healthy lineage could have emerged.

A major cause of the weakened physical condition was the occurrence of anchylostoma, also known as 'hookworm disease' (Lampe 1927). The labourers were used to 'doing their business' outside in the field and worked barefoot in the same (often muddy) soil. This made them susceptible to this disease (Bhagwanbali 2010, p. 138; Choenni 2016, pp. 420–421). Fokken (2018, p. 184) points out that labourers did their work barefoot because wearing shoes was a luxury at the time. That health was a continuous concern is also reflected in the various causes of death, including various worm diseases, suicide, malaria, syphilis, drowning and dysentery (Bhagwanbali 2010; Choenni 2016). Some common diseases could be traced to working conditions or debilitation due to overwork, but lifestyle, nutrition, hygiene and physical constitution were also important conditioning factors.

Poor health had important economic consequences. A report published in 1926 at the request of the Dutch Chamber of Commerce states: 'The most widespread disease is anchylostoma or hookworm disease ... If the disease is neglected, the constitution is strongly affected and the body weakened. Where ± 90 pCt of Javanese and British Indians are infected, the relatively low labour performance of agricultural labourers needs no further explanation. The anchylostoma feed on the blood of the victims and cause a weakening and anaemia, which certainly reduce the patient's working capacity to two-thirds of normal' (De Cock Buning 1926, p. 22). Given this physical weakening and the disproportionately heavy tasks, it is questionable whether the formally agreed minimum wages were met (see next section).

The health situation also had direct effects on population growth. In a detailed description of the health of the Surinamese population, the then-government medical practitioner Lampe (1927) specified the distribution and condition of the population over the period 1925–1927. He reported that on 31 December 1925, out of a total population of 139,869, there were 32,533 British Indians of whom 18,066 were men and 14,467 women. Of these, 1973 men and 1653 women lived in Paramaribo and 16,093 men and 12,814 women lived in the district (Lampe 1927, p. 21).

In the period 1921–1925, the surplus (i.e. the balance of births and deaths per 1000 inhabitants) for Suriname as a whole was 12.9 and for Paramaribo 3.2 (Lampe 1927, p. 375). This means that the districts, where most of the British Indians and Javanese lived, had a considerable population surplus. Lampe (1927, p. 375) explains that in 1924–1925, the birth surplus among British Indians was 28.1 per 1000 people, while among Javanese it was 15.2 and among the rest of the population 13.5.

The population growth among Hindustanis was the highest of any group. According to Lampe (1927, p. 344), this was probably related to '…the better family life and family maintenance; to family formation at an early age and to the relatively lower incidence of venereal diseases'. The high birth rate of British Indians persisted until the 1970s, but the recovery of the sex ratio occurred slowly as shown in the graph below, which depicts the male-female ratio for the period 1922–1970 by ethnic group (M/F x 100; see Annex 3.1).

Chart 3.1 shows that the ratio of men to women for the whole country fluctuated around 100 for the period 1922–1970 (second line from the bottom), indicating a balanced sex ratio. But this ratio varied by ethnic group, although they all converged from an unequal situation to be balanced around 1970. Among Creoles, women were in the majority (the bottom line), while among Hindostanis and Javanese it was men (respectively the third and fourth lines from the bottom). Demographic equilibrium was reached only around 1970, i.e. half a century later. The sharpest convergence in the male-female ratio occurred among the Javanese, but their starting position was uneven due to the immigration of mainly men that lasted until 1940. The immigration of Hindostanis stopped in 1916. Their annual growth rate in the period 1922–1943 per 1000 population was 29, while that of Creoles was 16.9 and that of Javanese due to continuing immigration 33.7 (Lamur 1973, p. 145).

Of the approximately 34,304 British Indian immigrants in Suriname, one-third (approximately 11,333) went back. Emmer (1984, pp. 230–231) reports that 5404 people, including 2000 boys and girls, died in the period 1873–1916. It is not clear whether this figure includes workers who lost their lives during the voyage. There

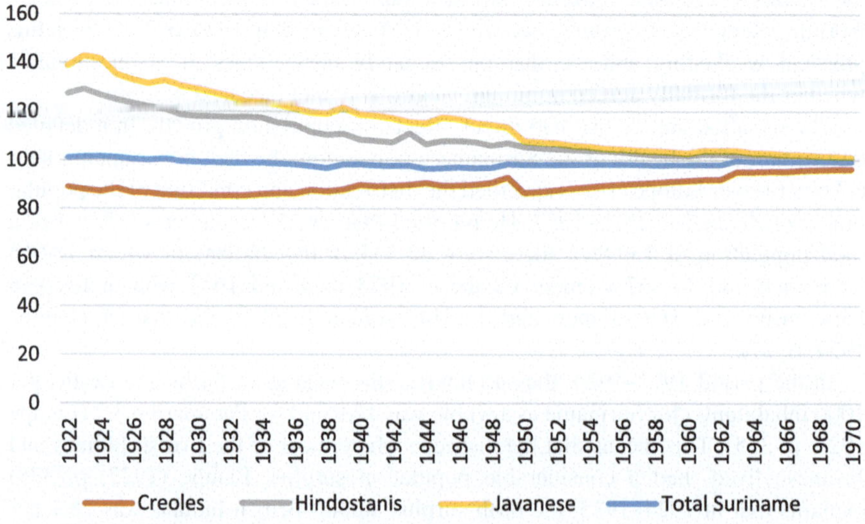

Chart 3.1 Sex ratio on the left (M/F x 100) by ethnic group, 1922–1970. (Source: Lamur 1973, pp. 178–179. Data were missing for the years 1961 and 1962 and have been filled in with averages for the years 1950–1970)

Table 3.1 Status of arrived British Indian immigrants in Suriname 1873–1916; numbers as percentage of arrivals

Arrived	34,304	100%
Deceased	5404	16%
Remained	28,900	84%
Returned	11,341	33%
Stayed in Suriname	17,559	51%
Potentially engaged in family formation	10,034	29%

were also 6399 births. These children cannot be compared to the working labourers. A breakdown of migrants arriving in Suriname is summarized in Table 3.1.

Of the 34,304 arrivals, about 5400, or one in six, died between 1873 and 1916. About 40% of the number of immigrants who arrived (11,341 out of 28,904 people) returned. In other words, almost one in two and a half went back. Assuming that the male-female ratio was 100 to 40, that left approximately 5016 women in a population of 17,563 immigrants who remained in Suriname. So potentially 5018 men and an equal number of women could engage in family formation, which adds up to 10,036.

Note that the above exercise is meant to give the idea that not all labourers could start a family. The totals pertain to the arrivals while the returnees include children born in Suriname. Moreover, the newcomers arrived over a period of 43 years, which distorts the numbers. The exercise cannot, and does not, have any claim on accuracy but is meant to give an idea of the scale of family formation during the indentured period. Note also that not everyone could have descendants: some of the adults may have had no partner, some of them died prematurely, and some were too sick and weak to make anything of life. So the proportion that could have potentially successful is a smaller percentage of the suggested maximum number of families.

The above outline implies that even with an equal sex ratio, not every Hindostani could have had a partner. The coming together of two people is highly dependent on the geographical concentration of the group, the barriers between the partners involved and what they have to offer each other. Thus, even with an equal sex ratio in a given area, differences in religion, castes or undesirable individual behaviours (alcohol consumption, abuse, orthodox views on culture and restriction of women's freedom, for example) can prevent the formation of families. Thus, a numerically equal sex ratio does not yet mean a social match for family formation. Given the wide geographical dispersion and long-term unequal sex ratio, family formation among British Indians must have been limited for a long time.

3.4 Wages

Wages are meant to make ends meet. When wages are too low, poverty ensues, and sometimes resistance or withdrawal. Negative treatment by plantation authorities also affected the commitment to plantation work negatively. From this perspective, some historians explain that former slaves had developed an aversion to working on the plantations and turned their backs on them (van Lier 1977). Since the remuneration of labour was a crucial factor for labour continuity and work experience, it is important to look at this issue more closely.

The wage for adult men was 60 ct. per day, and for women and boys and girls between the ages of 10 and 15 it was 40 ct., 'leaving open the possibility of paying a higher daily wage' (Snellen 1933, p. 53). The labourers enjoyed free housing and medical treatment. On re-indenture, they received 100 guilders, roughly equivalent to 166 days' minimum wage of an adult male worker. Labourers also received that amount when they waived the free return trip to British India and settled as peasants in Suriname.

According to De Klerk (1953, p. 131), the remuneration of labour on the plantations met 'the minimum requirements for a dignified existence'. This view is shared by earlier and later writers such as Snellen (1933), Emmer (1984) and Choenni (2016). De Klerk reported that several parties, such as the agent-general Van Drimmelen, the Suriname Commission (1911) and the British rapporteurs McNeill and Chimman Lal (1915), criticised the level of wages, but he maintained that the wage level was reasonable (1953, p. 135, footnote 86). I let him speak at length because this view contrasts with my argument:

> ... if one considers that the wage scale was the one in force in the colony; their (the labourers', RG) accommodation and medical treatment were free of charge; they were exempt from stamp duties on all documents relating to immigration and from personal taxes; their necessities of life were low and the cost of homegrown food low; moreover, they were granted a premium in the event of being re-indentured and in the event of relinquishment on free return ..., then one must admit that, all in all, their financial position was not as unsatisfactory as the low wage scale, taken by itself, might suggest. The British Indian immigrants therefore always managed to save not inconsiderable amounts of money, which they invested in silver and gold coins, jewellery, expensive clothes, plots of land or cattle, or deposited at the Colonial Postal Savings Bank. With the experience which they had gained as contractors of the labour, climate, institutions and language of the colony, these assets formed a solid basis, on which they could work their way up to colonist without great risk. (De Klerk 1953, pp. 135–136)

The remuneration may look reasonable on paper, but detailed research shows that administrators thought differently about the practice (Hoefte 1998, pp. 118–126). Generally, the employer determines wage levels based on labour productivity: the higher the output per unit of time, the more the earnings. This productivity is in turn determined by the technological level of the production process, the heaviness of the tasks, i.e. the intensity of labour, which is directly related to physical condition, and experience or routine. However, the technological developments of sugarcane processing as described in the international literature are of little relevance because wage levels were contractually determined and were not related to labour productivity.

On the plantations, work was allocated based on implicit perceptions of race, gender and age (Fokken 2018, p. 151ff; Hoefte 1998, p. 114ff). The prevailing racial conceptions led to the strong tendency to assign the heavier tasks to Creole labourers, mostly ex-slaves, who possessed greater experience and were considered better for performing heavy tasks. 'These free Afro-Surinamese adult men were capable and allowed to move about freely, without passes, could sometimes bargain themselves into the better paying tasks and because of their command of Sranan Tongo—the lingua franca of the plantation—were much better equipped to make their demands and wishes heard' (Fokken 2018, p. 196). As well as heavy duties, light duties also existed. These were assigned to women and children, who were paid less. Hoefte (1998, p. 115) points out that planters thus introduced a gender-based division of labour.

Bhagwanbali (2010, p. 157) compiled a table based on the Colonial Reports and the Secret Resolutions of the Ministry of Colonies showing the average wage of a Hindostani immigrant over the period 1875–1891 (see Annex 3.2). That table shows that the 'able-bodied' labourers earned an average of f0.70 per work task and the non-'able-bodied' labourers earned f0.61, with the range being f0.66–f0.73 and f0.43–f0.61, respectively. The small variation range indicates that no major fluctuation occurred, especially considering the fact that in 11 out of the 16 years, the 'able-bodied' labourers earned more than 70 ct. and the earnings of the non-'able-bodied' labourers were 60 ct. or more in 13 out of the 16 years. Bhagwanbali does not say whether the work tasks coincided with a day task, but assuming this was so, it is notable that the average wages were higher than the minimum daily wages of 60 and 40 ct. Since these figures are annual averages, it is not yet clear how earnings were spread over the year. It is also unclear whether these were Hindostani labourers under contract or free labourers (Photo 3.1).

Photo 3.1 Showing the payout of wages on a plantation. (The photo comes from SH-Dikland, SurinameHeritageGuide)

Given the unequal power relations between planters and labourers, the question may be raised as to whether or not wages were manipulated. Bhagwanbali (2010, Annex 4.3) presents a table for the same period (1875–1891), also based on the Colonial Reports and Secret Resolutions of the Ministry of Colonies, showing the total annual wages paid to Hindostani labourers, broken down by men and women. Since 52 Sundays are subtracted from the 365 calendar days as well as 32 Hindu holidays plus an unknown number of general holidays (set the latter category at ten), the number of working days comes to 271. This figure does not include days not worked due to illness, seasonal fluctuations and other causes. De Klerk (1953, p. 131), however, speaks of 313 working days.

This figure cannot be correct because after deducting the 52 Sundays, 10 general holidays and 32 religious holidays (for Hindus), 271 working days remain. In addition, there were absentee days, sick days and other days when no work was done. In 1892, Dr. Comins reported that contract labourers worked at least 183 days a year, free Asians 219 days and Creoles 265 days. A few decades later, McNeill and Lal (1915) arrived at 184 days worked. The annual number of days worked seems to have fluctuated over the years. Hoefte (1998, p. 125) reports that from 1897 onwards, male labourers under contract worked no more than 190 days a year and women a maximum of 150 days.

Each figure to calculate the daily wage is somewhat arbitrary because the exact number of days worked is not constant and probably varied with the type of plantation and the individual ability of the worker. For the sake of consistency, I join Hoefte in using 190 days for men and 150 days for women. If the annual income presented by Bhagwanbali is divided by these numbers, the minimum and maximum daily wages for men and women rise to f0.56 and f0.78 and f0.47 and f0.56, respectively. Among men, the daily wage was f0.70 or higher in 11 out of 17 years, and among women it was f0.50 or higher in 8 out of 17 years, in both cases higher than the contractually agreed minimum wage.

Hoefte (1998, p. 122), based on the report of McNeill and Lal (1915), works out the weekly wages on sugar and other plantations in the period 1908–1912. Divided by six working days, the average daily wages for men and women on the sugar plantations come to 46 and 27 ct., and on the other plantations to 48 and 28 ct. This is a considerable difference from the amounts mentioned above. It is unlikely that this difference was due to the number of days worked because Lal and McNeill's statement does not differ much from Hoefte's. The most credible explanation is given by McNeill and Lal (1915, p. 155) themselves when they state that wages had fallen (too far). But this explanation presupposed a market operation that applied only to free labourers and not to contract labourers, unless planters managed to decline or reduce payments.

Both labourers and planters had the ability formally to file complaints about each other, but this was a hopelessly unequal fight because in practice, planters handed out punishments. In 1863, the power to punish came into the hands of the district commissioner and later the district judge, to be finally abolished in 1947. According to De Klerk (1953, p. 123ff), the Poenale Sanctie, a relic of the slave era, gave the planter the right to punish the worker. This reading is contradicted by Bhagwanbali, who argued that corporal punishment, after being abolished, was reintroduced in 1875 (Bhagwanbali 2010, p. 190). How often this punishment was put into practice is unclear, but it was notable that British consuls in Suriname in particular repeatedly protested against the use of corporal punishment, especially flogging and cuffing (Bhagwanbali 2010; Fokken 2018).

The opportunity for labourers to lodge complaints was limited. Bhagwanbali (2010) points to the language and cultural barrier and the social circle to which planters and district commissioners belonged. These stifled the desire to file complaints against authorities. The only benevolent body to which labourers could turn was the British consul, who was generally critical of the treatment of British Indian labourers. Labourers needed a pass to leave the plantation, which was often refused if they wanted to file a complaint against the planter or a district commissioner. Moreover, complaining labourers were harassed by being employed to cut down thorn bushes or clear latrines, for example. Bhagwanbali (2010, p. 174) stated: 'The right to complain against injustice depended on the benevolence of the planter. And those who did not comply risked irrevocable punishment. Many did not dare to take this risk and then refrained from filing a complaint.' Such effects, stemming from the plantation hierarchy and the intention to control labourers, have also been addressed by Hoefte (1998) and Fokken (2018).

Planters also had complaints about the labourers, but they had more opportunities to express their dissatisfaction. Their main complaints were work refusal, failure to complete work tasks and laziness. Illness, non-attendance at the place of work and heavy tasks not completed fell under these categories and were punished by whipping, fines and imprisonment. Bhagwanbali (2010, p. 191) works out that in the period 1880–1883, the proportion of migrants sentenced to fines and imprisonment ranged from 31% to 45% of the number of contract labourers present in the colony. Judging from Bhagwanbali's figures, the proportion sentenced to prison ranged from 13% to 19%, leaving aside the fact that prisoners were also flogged and beaten (Bhagwanbali 2010, pp. 191–193). The length of imprisonment is not given, but the high percentage of prisoners reflects the fact that discontent among the planters was high and that they were willing to withdraw a lot of labour from the plantations. Fokken (2018) and Hoefte (1998) additionally point out that the very people with whom the labourers came into daily contact, the supervisors, were the ones about whom they complained.

Fokken (2018, p. 157) compiled a table of the most common complaints by labourers about plantation authorities during the period 1874–1919. Most of these related to underpayment, overwork and mistreatment. The supervisors possessed a high degree of discretionary powers and could afford to punish and harass. These circumstances on the plantations were addressed in several official reports, most notably by magistrates McNeill and Lal (1915), the British researcher Comins (1892) and the British consuls. But despite the many 'patrons' of the labourers—the colonial government, the agent-general, the district commissioners, the medical inspectors and the various British consuls (De Klerk 1953, pp. 134–135)—they did not succeed in structurally improving the labour situation on the plantations.

Bhagwanbali (2010, pp. 152–159), basing himself on reports by British consuls, agents-general and district commissioners, gives numerous examples showing that wages were not paid and that planters were only scantily complying with agreed obligations. The most concerned authorities knew about this, wrote letters to their superiors and obediently submitted their annual reports. Foreign inspectors of working conditions also denounced this situation. In some cases, labourers had to live on 'rubbish, leaves and weeds, which they picked and sometimes ate raw; others lived on petty (fruit and vegetable) theft' (Bhagwanbali 2010, p. 156). This was offset by a rise in the cost of living. The combination of low monetary income and the rising cost of living maintenance led to abject poverty among some of the labourers. It was not until 1920 that wages were formally raised, motivated mainly by the rise in the cost of living (see also Hoefte 1998, pp. 120–123).

Poverty and poor nutrition in turn had the effect of further worsening health. In 1881, for example, Agent-General Brandon reported to the governor that immigrants were receiving less than the agreed daily wage because it was linked to tasks. As a result, the labourer could not possibly earn enough '…for his maintenance; consequently his constitution deteriorates, and anaemia and susceptibility to indispositions develop' (quoted in Bhagwanbali 2010, p. 144). This situation may explain the high 'work absenteeism' (laziness) and the astonishingly low income of some of the labourers.

The exploitation of labour, individual harassment, curtailment of their freedoms and molestation of women led to both passive and active resistance. The first took different forms and ranged from individual sabotage to running away, arson, vandalism and violence against the authorities, and sometimes even killing a planter. The second took the form of rebellions, often outbursts of accumulated frustration. In these cases, the police and army were called in and several people were killed. Plantation resistance has received ample attention in the literature (Choenni 2016, p. 584ff; Hira 1982; Hoefte 1998, p. X), but mostly these conflicts have been approached as conflicts that occurred mainly on sugar plantations and not as responses to the ongoing exploitation of labourers on all plantations. The emphasis on sugar plantations may be justified but is insufficiently grounded in the literature.

Labourers sometimes ran away, known as 'desertion' and described as staying away from the plantation for 48 h without notice (Bhagwanbali 2010, p. 179). Desertion was a cost to the planter who had contributed to the expenditure on the workers maintenance and facilities on the plantation. Desertion also led to a reduction in available labour on the plantation. Although migrants risked fines and prison sentences, desertion was so common that surveillance on plantations was stepped up, river patrols were instituted in Suriname's border districts and bounty hunters were deployed to bring back runaway migrants (Bhagwanbali 2010, pp. 180–81; Emmer 1984, p. 233; Fokken 2018, p. 335). Desertion was understood in labour jargon as work absenteeism. It meant that both those who had run away and those who had to serve a prison sentence had to make up for their time not worked.

According to data provided by Fokken (2018, Annex 3.5), desertion was not a continuous phenomenon. The number of Hindostani deserters ranged from 1 to 28 per year. There were no clear peak years. It is notable, however, that no desertions occurred after 1902. In contrast, Emmer (1984, p. 233) cites figures of numbers of deserters up to and including 1916, totalling 327 cases. The outliers in his figure series were 28 runaways in 1892 and 120 in the period 1884–1890, followed by a gradual decline until the twentieth century. Then there was an outlier of 18 runaways in 1905 and increases in the years 1913, 1914 and 1916 to 17, 13 and 12 cases, respectively. The trend in runaways was thus declining, although there were periods of brief upturns. Some of these runaways returned (a total of 122), but the numbers were small. The largest numbers of turn-ins were in 1880 and 1914 with 12 and 16 cases, respectively. Most of the runaways by far were men (286 out of 327 cases), and the number of boys and girls combined was 15 over the whole period.

Explaining the decline in the runaways trend is not easy. It may be that working conditions on the plantations had improved, but there are no indications of this. It is also possible that labourers bought themselves free to settle elsewhere or simply ran away without bothering about the contract stipulations. The possibility of this was increased by the fact that after 1895, the government switched to a colonisation policy and made land available for settlement. As with differences in wages, the problem arises that a phenomenon can change over time and acquire a different meaning. As a result, the explanatory value of a factor is not constant. Allen (1999), in an analysis of plantations in Mauritius, believes that historians often overlook this point. In the historiography of Surinamese Hindostanis, this is sometimes the case, resulting in a lack of nuance. This is true, for example, of De Klerk (1953), who paints one picture for the entire immigration period, and Bhagwanbali (2010), whose data cover the last quarter of the nineteenth century while he generalises it to the end of the immigration period.

3.5 Savings

Migrants returned to British India with valuable items consisting of jewellery, coins and cash. The question is how the migrants' savings should be understood (Photo 3.2).

The photo above shows a woman in good shape. Part of her outfit was the jewellery that most likely had been accumulated in Suriname. It indicates that the labourers could save and spend or invest in jewellery. This accumulation was not restricted to those who stayed behind but included returnees. Data collected by Choenni (2016, p. 509; see Annex 3.4) show that such valuable items were substantial. The amounts of money, coins and jewellery added up and divided by the number of passengers, including children, ranged from f127.43 to f425.35. From these amounts, some authors have concluded that remuneration on the plantations was not so bad (Choenni 2016; De Klerk 1953; Snellen 1933). The underlying reasoning presupposes that saving was possible thanks to wages. For example, Choenni (2016, p. 525) writes: 'Most probably, many were able to save about half of their wages. If we assume one contract period (five years) and 180 working days per year, it is plausible that they were able to save on average about 25 ct. per working day.'

Photo 3.2 British Indian woman in holiday attire. (Source: Fotostudio 'Augusta Curiel'. NMVW-collectie (wereldculturen.nl))

Bhagwanbali (2010, p. 166) disputes this statement, at least part of it, on the basis of an eyewitness in 1886 who claimed that many coins had been brought back from British India. The witness read this from the fact that the coins were British Indian coins.

However, this explanation is not convincing as the coins could have been 'ordered' later. From photographs of arriving immigrants, it can be seen that they wore little jewellery, while labourers living in Suriname were adorned with it. This emerges in a comparison of Photos 2.2 and 3.2. The jewellery was mainly silver and arguably the money to pay for the jewellery must have been earned in Suriname. Other authors come to a similar conclusion (Choenni 2016, pp. 525–526; De Klerk 1953).

But not everyone went back 'rich'. Bhagwanbali (2010, pp. 167–168) concludes, on the basis of passengers from five ships in four different years, that the picture was less rosy. His calculation contains some ambiguities and therefore a recalculation is given below.

The amounts are quite high, but since the purchasing power of the money is unknown, the significance of these amounts is better indicated by the men's agreed minimum wage. The lowest per capita amount in Table 3.2 (last row) is 209, and dividing by the daily wage of 60 ct. gives a total number of 348 working days. That is almost twice the annual income of the workers!

Table 3.2 Number of departures and values taken by passengers of five ships, 1878–1897

Year	1878	1878	1884	1895	1897
Name of the ship	*Philosopher*	*Sint Kilda*	*Silhet*	*Grecian*	*Foyle*
Number of migrants	476	333	537	605	213
Number with money and jewellery (values)	221	112	215	198	105
Number without value	255	221	322	407	108
Taken value in *f*	46,304	25,466	123,527	105,189	30,534
Per capita	209	277	574	531	291

Source: Recalculated from Bhagwanbali (2010, p. 167)

Also striking is the fact that over all these years, more than half of the passengers did not take any such valuable items. This does not mean that they did not possess any. The migrants also transferred money to British India earlier. But that would mean that they had transferred everything and returned with 'empty hands'. Moreover, it is also unclear over how many contract periods these savings were collected. The amounts are averages of labourers with different contract periods, with and without partners, and therefore rough indications.

An observer of the time, despite his recognition that the labourers had shown great thrift, disputed that the savings had been amassed from wages (Wesenhagen 1886). He explained the origin of savings from various other activities: 'Part of these so-called savings are therefore contributed from the sale of bananas and fruits of the earth, and occasionally from the profits of a successful goat farm, which is peculiar to them. For the rest, they are largely derived from a few coolie chiefs, who often abuse their authority to engage in all kinds of gagging, while occasionally the origin of these funds must also be sought in the advantageous exploitation of married coolie women by their husbands and of unmarried ones by themselves' (Wesenhagen 1886, pp. 49–50).

It is very likely that these sources of savings shifted in importance over time. Snellen (1933, p. 69) states: 'During the first 20 years, when no significant colonisation took place, these possessions arose exclusively from saved plantation wages; in later years, what was earned through work on independent farms or in trade was added; but already with the first shipments, amounts of f100 to f200 per head of the departed, including children, were taken away, so that with each departure large amounts were withdrawn from circulation in the colony.' Wesenhagen, writing in 1886, refers to entrepreneurship as sources of savings, while according to Snellen, savings were set aside from wages during the first 20 years. Note, however, that savings were not only an effect of income earned but also of (family) consumption, especially the number of mouths to be fed.

Choenni (2016, p. 464) asks why labourers would enter into a re-indenture when they could earn more as free labourers. The answer to that question is that with a re-indenture, labourers were assured of a premium, and free labourers had no certain claim to a fixed income. For the planter, labour costs were lower under a re-indenture, partly because there were no supply costs in the second and later term and the labourer was more productive due to the experience gained (Bhagwanbali 2010, pp. 147–150). Overall, the amounts saved from both wage income and other sources were, on average, higher in households with partners than in single-person households. Families received several premiums when leaving or re-indenturing. If the worker had a partner who also worked, the premium was twice as high.

Self-employed labourers' enterprises were another source of savings. These were mainly enterprises in agriculture and commerce, where land was sometimes bought and sometimes obtained from the government as rental land. De Klerk (1953, p. 161) points out that when renting domain land in Suriname to practise horticulture, the immigrant did not lose their right to free return. Some authors have also pointed out the existence of these different sources of income (Burger 1928; Snellen 1933). Therefore, it is wrong to say that savings only came from wages (Bhagwanbali 2010; Choenni 2016). It is overlooked that there were multiple sources of income and therefore savings could not be directly related to wages earned.

3.6 Own Cultivation

The fact that labourers, especially in agriculture, grow their own food does not always stem from poverty. It is often a habit of people still in, or close to, agriculture to grow their own food; usually growing crops for their own consumption is part of a lifestyle. To grow their own produce, the labourer must have access to land. This can be done by a large landowner, such as the plantation owner, providing the peasant with a piece of land in exchange for certain services. Sometimes the labourer bought a piece of land or obtained or rented it from the government. All four forms of access to land occurred in Suriname (Choenni 2016; Fokken 2018; Bosz and Albrecht 1954). The title to the land (ownership, long lease, leasehold and the like), the quality of the plot, the fertility of the land, the presence of infrastructure (roads, water outlet and supply), the location in relation to markets and, above all, the knowledge of crop cultivation helped to determine what and how much the labourer produced. On the plantations, growing crops was not intended to increase the worker's production but to secure the availability of wage earners and reduce labour costs.

Even back in the slave era, planters allocated their slaves, some of whom were surplus, a piece of land on the plantation to grow their own food (Van Stipriaan 1993, pp. 350–357). Through this arrangement, they did not have to maintain slaves, while the availability of labour was secured. The planters thus saved on the maintenance costs of the slaves. When they had to employ British Indian wage labourers this disposition of labour was contractually assured. After the contract expired, the labourers, who by now had been trained and become experienced and thus achieved higher productivity, could renew the employment contract. But this did not always happen. Sometimes labourers returned to British India and sometimes they settled outside the plantations. For planters, that was a loss of investment. This led to two strategies to secure labour.

The first was to provide contract labourers with a piece of land on which they could grow their own food such as rice and vegetables, as well as raising poultry, cows and goats. This was already the case in Suriname during slavery, when free 'slaves' were allowed to exploit a piece of land for board and lodging (Bosz and Albrecht 1954). British Indian indentured labourers were also granted these facilities early on, a system that lasted several decades (Comins 1892; McNeill & Lal 1915; Snellen 1933). For the labourers under contract in particular, this could mean a lower wage than agreed. The exact terms and practice of this access to land remain to be explored in detail, but generally this provision is known as a strategy to reduce labour costs (Goodman & Redclift 1981; Mintz 1983). This was a godsend for the planter because the alternative was more expensive: the free labourers possessed better bargaining power and made higher wage demands (Hoefte 1998). Because the costs of indentured labourers continued, regardless of whether they were assigned tasks or not, the free labourers were often employed only in peak periods.

The second binding strategy was to allocate land to the free labourers in exchange for labour services in peak periods. Snellen (1933, pp. 67–68) describes this practice as follows: 'The provision of land for growing rice and dry crops up to an area of one hectare strongly encouraged the free settlement of these people. In return, these labourers took on the obligation to work on the plantation for a set number of days a week or daily for certain activities, such as coffee picking. Not all settlements were in a position to set up such establishments, as suitable land was not always available in sufficient quantities for this purpose. Where this was the case, it was generally considered a sound means of keeping good families, preferably with children, on the plantations. Yet that measure only prevented the migration of British Indians from the plantation in a few cases; the majority of the British Indians sought a completely independent existence, either on a piece of land or in the city.'

These strategies were not followed by all planters. Sometimes the cultivation of their own subsistence was banned—for instance, because no land was available. However, labourers rarely bothered about the restrictions. They kept cows, goats, pigs and chickens, grew coffee, bananas, bacoven, rice and maize. Comins observed such developments as early as 1891, which indicates a strong agricultural continuity during the phase of proletarianisation. Similar findings have also been reported by Hoefte (1998), Bhagwanbali (2010), Choenni (2016) and Fokken (2018). Based on photographs and descriptions, Fokken (2018, p. 176ff) shows that labourers grew bananas, mango, sugarcane and maize both on plantations and in kitchen gardens; they also kept animals. Even if planters did not allow such activities, there could still be space for them outside the plantations on nearby land.

Growing crops for one's consumption and keeping cattle indicate a (semi-)agricultural subsistence. The deskilling that Braverman (1974) pointed at could not be complete for these labourers because access to the land was not closed and migrants used their knowledge of growing agricultural crops and means of subsistence. Thus, starting an agricultural business was not only enforced by poverty or the planter's desire to reduce labour costs. The worker also had his own motive, namely to make a profit because this contributed to his savings. The latter meant that any surplus produce had to be sold.

What were the outlets? The surplus of produce was sold on the plantations with large numbers of labourers, at the district headquarters and in the city of Paramaribo. During the week, labourers were at work on the plantations and therefore sales had to take place mainly on Sundays and during holidays. Not much is known about this. McNeill and Lal (1915, p. 165) recounted that 'on Sundays labourers flock to Paramaribo by boat both to the market and to meet friends and acquaintances. Visits are also paid to other plantations or settlements'. Since not all plantations offered the same opportunities and the labour regime also varied, it can be assumed that the opportunity to sell produce was unequal. The location of the plantation, the product grown and the plantation regime (including the benevolence of the individual planter) largely determined how the labourer performed economically.

3.7 Conclusions

This chapter has addressed the experiences of plantation labourers as it was during this period that the conditions were created for later settlement of the peasants. These experiences included not only the labour process but also issues such as wages, treatment, savings and the development of entrepreneurship. Each of these factors increased the diversity among labourers in terms of resources, experiences and opportunities to start as a peasant. The experiences are mainly based on the sugar plantations, which is a limitation because work and labour regimes were more favourable on other plantations. Thus, the labourers did not form a homogeneous population but are treated as similar.

Two categories of factors highlight the differences: factors that belonged to the plantation and those that characterised the labourers. The first category comprised the product grown and the plantation regime, including the individual planter's character. The latter also varied over time as significant improvements had taken place in the plantation system. The second category of factors consisted of the physical condition of the worker, the type of household (family or not), the size of the household, and the sex and age of those involved. This variety of factors makes it impossible to generalise about the labourers.

The heterogeneity of labourers meant that the experiences gained were not the same for everyone. For example, the earnings (from labour, entrepreneurship and premiums) were affected by whether the labourer was single or had a family. Satisfactory earnings were also offset by unpleasant experiences such as flogging, cuffing, imprisonment, fines, bullying and other forms of molestation. It is unlikely that labourers who were subject to these experiences opted for a re-indenture. Therefore, a link between the sharp decline in desertion in the last quarter of the nineteenth century and the opportunity to establish themselves as peasants is very plausible. The differentiation of labourers was thus determined not only by characteristics of the labour process but also by what the labourers experienced. It is therefore noteworthy that this dark side of labour life is mentioned in passing in historical research.

The transformation to wage labour, the proletarianisation, showed some specific characteristics. It was for a short time, 5 years, unless those involved opted for a re-indenture, whereas in the normal process of proletarianisation, the transformation to wage labour is permanent. Also special was the fact that the labourers had the option of planting their own food on land located on or near the plantation and selling any surplus. Therefore, the labourers were 'proto-entrepreneurs' and not totally dependent on wages because they controlled part of their food. This enabled them to maintain their agricultural human capital. Thus, the transformation from rural people in British India to plantation labourers in Suriname was not only temporary but also incomplete. The British Indian labourer remained partly a peasant on the plantation.

References

Allen R (1999) Slaves, freedman, and indentured laborers in colonial Mauritius. Cambridge University Press, Cambridge

Bhagwanbali R (2010) Hindoestaanse migranten onder het indentured labour systeem naar Suriname, 1873–1916. De nieuwe awatar van slavernij. Amrit, Den Haag

Bosz AJAQ, Albrecht AJ (1954) Drie Eeuwen Grondpolitiek in Suriname. Van Gorcum & Comp, Assen

Braverman H (1974) Labor and monopoly capital: The degradation of work in the twentieth century. Monthly Review Press, New York

Burger JW (1928) Vergelijkend Overzicht van de Immigratie en Blijvende Vestiging van Javanen en Brits-Indiërs in Suriname. De Economist 77(1):422–445

Choenni C (2016) Hindostaanse contractarbeiders 1873–1920. Van India naar de plantages in Suriname. LM Publishers, Volendam

Comins DWD (1892) Note on emigration from the East Indies to Surinam or Dutch Guyana. Printed at the Bengal Secretariat Press, Calcutta

De Cock Buning W (1926) Rapport over Suriname, uitgebracht aan de Vereeniging van Kamers van Koophandel en Fabrieken in Nederland

De Klerk CJM (1953) De Immigratie der Hindostanen. Urbi et Orbi, Amsterdam

Emmer P (1984) Een goede keus? De migratie van vrouwelijke contractarbeiders van Brits-Indië naar Suriname. OSO, Tijdschrift voor Surinaamse Taalkunde, Letterkunde en Geschiedenis 3(2):213–234

Emmer P (1985a) De werving van kontraktarbeiders in India voor tewerkstelling in Suriname, 1872–1916. OSO, Tijdschrift voor Surinaamse Taalkunde, Letterkunde en Geschiedenis 4(2):147–157

Emmer P (1985b) The great escape: The migration of female indentured servants from British India to Surinam, 1873–1916. In: Richardson D (ed) Abolition and its aftermath: The historical context, 1790–1916. Frank Cass, London, pp 245–266

Fokken M (2018) Beyond being koelies and kantráki. Constructing Hindostani identities in Suriname in the era of indenture 1873–1921. Doctoral dissertation, University of Groningen, Groningen

Goodman D, Redclift M (1981) From peasant to proletarians: Capitalist development and agrarian transitions. Basil Blackwell, Oxford

Hira S (1982) Van Priari tot en met De Kom: de geschiedenis van het verzet in Suriname 1630–1940. Futile, Rotterdam

Hoefte R (1990) De betovering verbroken. De migratie van Javanen naar Suriname en het Rapport Van Vleuten 1909. Foris Publications, Dordrecht
Hoefte R (1998) In place of slavery. A social history of British Indian and Javanese laborers in Suriname. University Press of Florida, Gainesville
Kumar A (2017) Coolies of the Empire. Indentured Indians in the sugar colonies 1830–1920. Cambridge University Press, Cambridge
Lampe PHJ (1927) Suriname. Sociaal–Hygiënische Beschouwingen. De Bussy, Amsterdam
Lamur H (1973) The demographic evolution of Surinam 1920–1970: A socio-demographic analysis. Martinus Nijhoff, The Hague
McNeill J, Lal C (1915) Report to the government of India on the conditions of Indian immigrants in four British colonies and Surinam, Part II. London
Mintz S (1983) Reflections on Caribbean peasantries. New West Indian guide 57(1–2):1–17
Snellen E (1933) De Aanvoer van arbeiders voor den landbouw in Suriname. Doctoral dissertation, University of Wageningen
Van Lier R (1977) Samenleving in een grensgebied: Een sociaal–historische studie van Suriname. Emmering, Amsterdam
Van Stipriaan A (1993) Surinaams Contrast. Roofbouw en overleven in een Caraïbische plantage–economie, 1750–1863. Brill, Leiden
Wesenhagen AC (1886) De middelen door de Staten–Generaal aangegeven tot behoud van der kolonie Suriname van meer nabij beschouwd. Nijgh & Van Ditmar, Rotterdam

Chapter 4
An Embryonic Ethnic Community

Abstract This chapter deals with the emergence of an embryonic ethnic community on the plantations. While the workers' freedom of movement on the plantations was limited, there was a degree of social life where the workers lived together for at least five years. To that end, Sect. 4.2 details the relatively small numbers of workers on the mostly isolated plantations. Section 4.3 discusses the breakdown of the caste system and the position of women, i.e. the breakdown of traditional hierarchies. Section 4.4 discusses some aspects of social life reflecting this increased equality—in particular, the adaptation of familiar institutions such as religion. Section 4.5 addresses the specifics of community formation, particularly the *dipuá* and *jaháji* relations and expressions of nostalgia and violence. Section 4.6 addresses interracial relations as they might have played a role in community formation. The plantations represented a community where numerous customs that had been taken along were revived or took firm shape in institutions such as religious rituals and festivals.

Keywords Cast system · Women · Ethnic food · Grief · Festivities · Interracial relations

4.1 Introduction

In the depots in Calcutta and during the voyage to Suriname, *dipuá* and *jaháji* relationships arose that functioned as assumed family relationships. In addition, marriages were concluded in the depots in Calcutta and Paramaribo. Also registered were the children who were born from existing and new relationships. The collection of workers were thus partly related by blood relationships or assumed family relationships. Although not all migrants were relatives, they did form an 'interactive quasigroup', i.e. people who were to some extent familiar with each other's routines and habits but did not consider themselves to be part of the same group (Mayer 1966, p. 97). Fenton (2003, p. 115) calls this latent cultural affinity 'tacit ethnicity'. The cultural and emotional distance was probably not very great, although the

R. Gowricharn, *The Girmitiya Peasants in Suriname*,
https://doi.org/10.1007/978-3-031-67961-2_4

labourers were aware that they came from different castes or from different regions and ethnic and religious communities.

While the workers' freedom of movement on the plantations was limited, there was a degree of social life where the workers lived together for at least 5 years, a period considerably longer than the stay in the depots and the journey from Calcutta to the plantations in Suriname. The plantations were thus not only a production unit but also represented a community where numerous customs that had been taken along were revived or took firm shape in institutions such as religious rituals and festivals. The large number of free holidays for Hindus and Muslims indicates a high level of religious institutionalisation (Choenni 2016, pp. 705–706; De Klerk 1953, p. 131). In addition, marriages and funerals are good examples of such institutionalisation. Social entertainment, mutual help in kitchen gardens, domestic help, and religious and other gatherings are also moments that must have determined internal institutionalisation (Bhagwanbali 2010; Fokken 2018, p. III).

The Girmitiyas suffered from grief and nostalgia. According to several authors, these emotions gave their ethnic ties a specific character (Bhagwanbali 2010; Majumder 2010; Narayan 2017). The emotions were manifested in musical expressions and lyrics of songs that mainly concerned nostalgia, homesickness and longing. A second feature of their social life at the plantations consisted of expressions of violence, excessive use of alcohol and suicide. These two categories of manifestations would have long characterised British Indians throughout the Girmitya diaspora (Choenni 2016; Jayawardena 1963; Lal 2000).

In regard to interracial relationships, in Suriname these have developed between different ethnic groups on a largely isolated plantation. Indications that racial blending occurred on a significant scale are few and far between (Hoefte 1998). Apart from racial blending, there is also the possibility that cultural blending may have occurred, e.g. of language, food, dress and musical expressions. But whether this occurred, and on a scale that affected ethnic group formation, is difficult to say. Nevertheless, it is a tendency that needs to be explored to understand the incipient ethnic group formation.

Ethnic group formation is important because productive forces were locked up in the group. In this, I distinguish two forms that Geertz (1983/2000) had labelled 'local knowledge': embodied and non-embodied productive forces. Embodied productive forces consist of the knowledge that each member of the community carried, often knowledge that the whole community also possessed at the time. For agriculture, that human capital included knowledge of the weather, soil (types), crops, planting, maintenance, harvesting and processing of the product. This shared human capital allowed neighbours and relatives to be involved in the production process.

The non-embodied productive forces also required embodied knowledge but included disembodied material such as work tools, seeds, storerooms, animals and the like. The embodied and non-embodied productive forces together constitute agrarian capital and cannot be separated from the ethnic community—not only because the productive forces were embedded in the ethnic group but also because they could only be mobilised through ethnic ties.

This chapter argues that the ethnic group formation of Girmitiyas was partly a continuation of the ties formed earlier in the depots in Calcutta and during the journey to Suriname. Ethnic group development was determined by the institutionalisation of religious and social practices such as rituals, mutual aid and entertainment, as well as the maintenance and consolidation of cultural practices. The ethnic group formation is important because it contained the productive forces important for the maintenance and expansion of agrarian human capital. The ethnic community thus represented a moral economy that had its roots in the mutual recognition of incentives, obligations and trust created through interaction and transactions over time (Carrier 2018).

The next section discusses the relatively small numbers of workers on the mostly isolated plantations. Section 4.3 discusses the breakdown of the caste system and the position of women, i.e. the breakdown of traditional hierarchies. Section 4.4 discusses some aspects of social life reflecting this increased equality—in particular, the adaptation of familiar institutions such as religion and a variety of occupations that workers pursued alongside their official work. Section 4.5 addresses the specifics of community formation, particularly expressions of nostalgia and violence. This led to a strongly one-sided self-image that mainly focused on the suffering of the workers. Section 4.6 addresses interracial relations as they might have played a role in community formation. I conclude with some findings about the limited ethnic group formation on the plantations.

4.2 The Distribution of Workers

The labourers usually arrived at Paramaribo. Sometimes they were disembarked at other places such as Nieuw Amsterdam and Nieuw Nickerie. This depended on the limited depth of the rivers and the type of ship. Occasionally, a ship had to wait in quarantine at sea because of fear of contagious diseases (De Klerk 1953, p. 79). But by far the majority of ships were unloaded in Paramaribo. There the workers were housed in the so-called 'coolie depot', a barrack where the migrants caught their breath from the long sea voyage for a few days and sometimes a few weeks, where the administrative formalities were completed (which could take 2–3 weeks), medical inspection took place and where sometimes marriages were concluded.

De Klerk (1953, p. 87) states: 'For the British Indians, who were already in the colony, the arrival of a new ship was always a festive event. In large numbers they crowded around and into the depot at certain, set hours.' This behaviour was prompted by several motives: the new arrivals could have brought messages from the native land, the contact made them feel more closely related to Mother India (Bharat Mata), and well-to-do British Indians also tried to buy a wife free from the unmarried women in return for payment of f156, two-fifths of the landing costs.

After staying at the coolie depot, the workers were assigned to plantations. De Klerk (1953, p. 117) reports that of the 410 workers who left Calcutta on the first ship, 11 died en route. The 399 remaining workers were mainly assigned to sugar plantations in the Commewijn district: plantation Zoelen 26 people, plantation De Resolutie 245, plantation Alliance 55, plantation Hooijland 53 and plantation Alkmaar 19 people. The total, incidentally, comes to 398 and not 410. Plantation De Resolutie stands out with an outlier of 245 people. If we leave out this plantation, the number of workers allocated to the plantations ranges from 19 to 55 people, including children. In this distribution, families were kept together, but *jahájis* could lose sight of each other. Although new contacts were made at the coolie depot, uncertainty remained over whether acquaintances would also be sent to the same plantations. In local speech, the term *kantráki* emerged to refer to the contract worker, the word referring to an '... insider perspective of those who not only saw their work, but also their personal life affected by the contract' (Fokken 2018, p. 150). It was not about a type of relationship like the *jaháji*.

Fokken lists the ten plantations with the largest number of British Indian workers under contract for the period 1875–1920. The size of the ten largest plantations changed every year as contracts expired, workers left or died, or fewer workers were needed. Sometimes they even dropped out of the top ten, while new plantations were added to the list. Altogether, there were 28 plantations. Over half of these were located in the Commewijne/Cottica district, and most plantations produced sugar-cane, cocoa and coffee as main crops. Fokken's data are summarised in this book as Annex 4.1, but it should be noted that not all columns add up to ten plantations. The data have been converted and summarised in the table below.

The table above shows seven periods over 45 years. Although the table provides a patchy insight, it does show trends—first, that the average number of workers under contract declined from 1890 onwards, with 1915 being a year of economic upswing. This pattern is also reflected in the mean and in the minimum and maximum number of workers (the spread). It may be that the number of labourers under contract declined because the number of available free and contracted labourers was sufficient to run the plantations and thus no new supply was needed. Secondly, it cannot be ruled out that the number of labourers under contract gradually decreased due to the decline in the number of plantations. And thirdly, there may have been some replacements by Javanese workers whose immigration had begun in 1890.

The third row in Table 4.1 shows that the average percentage under contract over the entire period was ten, including in the year 1885. The stable average rate of ten indicates that the ratio of the number of British Indian workers under contract to the total number of workers did not vary much. The diminishing number of workers

Table 4.1 Number of British Indian plantation labourers under contract on the ten largest plantations, 1875–1920

	1875	1880	1885	1890	1900	1915	1920
Total number (1)	1801	2257	3014	2835	551	1858	427
Average per plantation (2)	150	226	301	315	61	186	43
(2) as percentage of (1)	8	10	10	11	11	10	10
Spread per plantation	95–461	131–701	180–439	151–689	51–551	91–360	17–67

Source: Calculated from Fokken (2018), Annex 3.3

under contract indicates a decreased supply of new labourers and, therefore, a decreasing influence of newcomers on community formation on the plantations. The significance of *dipuá* and *jaháji* relations on the plantations thus declined.

Snellen (1933, p. 79) mentions that up until 1931, 11,623 people returned to British India, 140 of whom left at their own expense. With each returning ship, a number of elderly and destitute people were sent back because they were no longer productive for the community (Snellen 1933, p. 69). Smaller numbers of these returnees moved to other plantation colonies. This happened for two main reasons: the first was that they were part of the international flow of money-earning labour migrants (i.e. the migration culture); the second reason was that, for various reasons, some migrants could not settle in the homeland and returned to a colony disappointed (De Klerk 1953, p. 155ff). These small numbers of returnees were hardly significant for the development of later ethnic communities. However, they did provide a supply of news, products and the like from the homeland and the home culture.

4.3 The Cultural Equalisers

Apart from a small percentage difference of arriving and departing women, there are no notable differences in the demographic characteristics of the workers and those who went back to British India. Thus, the characteristics of the plantation population were not affected by arriving and departing workers. Given these preconditions, two processes on the plantations were important for the emergence of a new ethnic community: the breakdown of the caste system and the improved position of women. This is the process that has been referred to as the 'great equaliser' (Lal 2006, p. 371), which began in the depots of Calcutta, was maintained on the ships and continued on the plantations.

4.3.1 The Breakdown of the Caste System

Differences in caste were ignored on board the ships and on the plantations. Everyone was a passenger or a labourer and was treated that way. Thus, a breakdown of traditional hierarchies took place, starting in the depots, ships and plantations. While the breakdown on the ships took a few months, the social equalisation in the plantations lasted for a longer time. However, this did not mean that the caste system had no meaning outside working life. For the generation that came directly from British India, it can be assumed that the meaning of the caste system remained largely intact. But the conditions necessary to keep the system functioning were hardly present. The main missing factors were a sufficient number of women, a kitchen of their own and the size of the population on the plantations.

De Klerk (1953) cited the shortage of women as the main reason for the decline of the cash system. Members of (higher) castes were required to marry women of equal castes, but this was not always feasible. The months-long stay in the depots and on the transport ships had fostered mixed networks and marriages among the migrants. This was reinforced, says De Klerk (1953, p. 171), 'when there were enticing economic advantages attached to the union of widely separated castes'. The scarcity of women of equal caste was a structural constraint that could not be removed in a few years. Even if the caste element was waived, the absolute shortage of women was so great that it took practically half a century to redress the balance. It can be seen from Annex 4.1 that the scarcity of women was greatest during the period when labourers worked on the plantations.

The scarcity of women did not only erode differences among castes. Differences between Muslims and Hindus also gradually lost their significance. This was reflected in various acts, including marriages (Hassankhan 2016, p. 195). Such traditional barriers are sometimes explained by the scarcity of women (Choenni 2016), but this instrumental perspective leaves room for doubt, for several sources report that Muslims and Hindus mutually attended each other's festivals and supported each other in different ways (De Klerk 1953, p. 197; Fokken 2018, p. 178). Numerous narratives in Hindu circles also reveal that Muslim women suckled children of Hindus and vice versa when the biological mother did not produce enough breast milk, which was often blamed on hard work and poor health. This practice also created a semi-familial bond between two families.

Purity is an important element in the caste system. The concept does not refer to hygiene but to the avoidance of all kinds of contact between members of higher and lower castes. A distinctive element in this contact is the preparation of food and the use of kitchen utensils. Caste regulations were hardly taken into account on the ships (Kumar 2016), and it was no different on the plantations (Bhagwanbali 2010; Choenni 2016), thus rendering it impossible to maintain caste regulations. Further, the shared kitchen meant that workers had to take turns cooking and lost time

Jayawardena (1963, p. 19) described a similar situation in Guyana as follows: 'They divided the coolies into gangs, made them work next to each other and appointed persons to positions of authority irrespective of caste. They housed the coolies cheek by jowl in long barracks and accommodated them together in the hospital. Coolies shared the same water supply, bathed in the same canal and used the same latrines.' In Majumder (2010, p. 31), a labourer recounts that the upper castes in British India had their own wells. Such a separate social life, as the caste system commanded, was not possible in Suriname due to the small population and because the plantation regime did not allow it. Despite the breakdown of the caste system, there still remained a need for Brahmins who traditionally performed rituals.

On the plantations, there was a need not only for ceremonies at initiation rites, known as *muran sanskár* where babies were shaved bald, but also at marriages, deaths and other events. In addition, *kathas* were held, which were similar to a mass performed at home, which sometimes involved planting large and small bamboo sticks to which coloured pieces of textile were attached, known as *jhandis* (Fokken

2018, p. 187). All these rituals required pandits, who were usually members of higher castes. In some cases, members of the lower castes fulfilled the function of a pandit. For example, '[A]mong migrants living at the plantations, not only high caste men were found in the position of religious leader, but also some of lower caste status' (Fokken 2018, p. 186). It is unknown whether this 'ascendancy' occurred because members of higher castes were absent from the plantation in question, or because the acting pandit had a knowledge of rituals or for some other reason.

Majumber (2010, p. 37) points out that thanks to the presence of high castes, there was a continuity of Indian religious tradition. However, the example above shows that the requirement of caste origin was not an inexorable law: members of lower castes could also fulfil the role of pandit. But a pandit was required regardless of caste origin. Since most pandits were members of higher castes, their role in social life was certainly not played out. The continuous need for people who could perform rituals made it likely that changes of caste, as happened during registration and transport to Suriname, continued on the plantations. In other words, the plantations were not only an equaliser of hierarchy but also a place where new hierarchies emerged.

Until the first decades of the twentieth century, Hinduism consisted of a multitude of local religions originating from British India. This resulted in numerous religious communities, each centred on one or more god(s) from Hinduism. What is now known as 'Hinduism' previously consisted of a number of separate movements that did not share a common origin but comprised a collection of faiths that originated on Indian soil. Surinamese Hindus knew followers of, for example, Ram, Shiva, Vishnu, Hanoeman, Kali Mai, Durga, Parvati, Sarasvati, Laksmi and Siew. The exact status of these deities is debated as they are sometimes interpreted as incarnations of central deities (Brahma, Vishnu and Shiva) or as locally existing deities that were incorporated into the Hindu pantheon. The acknowledgement of all these deities, cults and currents also explains the relatively large number of days off given as holidays to Hindu labourers (see Choenni 2016, pp. 705–706).

In the second decade of the twentieth century, missionaries of the Arya Samaj came to the Caribbean, mostly people who not only carried out their religious mission but who also propagated Indian nationalism (Samaroo 2006). The Arya Samaj was formalised in the 1920s and the Sanatan Dharm also emerged in response. Muslims on the plantations experienced similar processes of religious affiliation and organisation (Hassankhan 2016). According to Fokken, this did not lead to a separation between Hindus and Muslims. 'Domestic religious ceremonies seem to have played a key role in bringing immigrants together, independent of whether or not they were Hindu or Muslim,' wrote Fokken (2018, p. 185). Mixing between Hindus and Muslims was seldom an issue. In all likelihood, the mixing, especially via unions, was greater than is usually assumed. Such far-reaching cultural mixing has also been reported for Bengal (Alexander et al. 2018).

4.3.2 The Position of Women

The role of women on the plantations is relevant to my argument insofar as it affects later settlement. That significance includes the family formation and related labour capacity and the level of savings they made possible. On these issues, Emmer (1986 1987) and Hoefte (1987a, b) engaged in a polemic. Emmer argued that the settlement of British Indian women in Suriname was a good choice. He based this on the lower mortality rate, the larger number of children and the higher wages they would earn in Suriname than in India. Although Emmer acknowledged that the population in India and Suriname cannot be compared, he persisted in his conclusion that settling in Suriname was a 'good choice'. However, he assumed that the differences in mortality and birth between the two countries had the same causes. Since populations and causes were different, his conclusion may be questioned.

What matters for my argument is not so much the emancipation of women—the central theme of the discussion—but their pay. Various sources reveal that women were grossly underpaid and therefore dependent on their partners. Emmer's (1984, p. 223) claim that women did not participate in plantation work because they took care of the household is based on an 1892 report by Comins. It cannot, therefore, have covered the entire immigration period that lasted till 1916. His rosy depiction that women could afford not to work is, if this occurred, not supported by any other source and probably did not occur on a large scale.

Relying on a colonial report, Hoefte (1987b, p. 122) writes in response: '(... the value of women was limited to their sexual quality. They had to improve the gender ratio and keep the workers satisfied. Since the women were on the plantations anyway, their labour power was used.' This answer is unsatisfactory. Why did women have to improve the gender ratio if the labour contract lasted 'just' 5 years? This was not the aim of recruiting workers. And if it was, why not agree on a ratio of 100 men to 100 women?

Not only was the ratio of men to women very unbalanced, but the distribution across plantations did not have to be proportional either. This meant that on some plantations there was a better, and on others a worse, ratio of men and women. The ratio was affected by early death, illness, relationships with plantation authorities and imprisonment (Choenni 2016). Incidentally, the scarcity of women only indicates a quantitative relationship and says little about the factors that actually led to pair formation. The fate of women on plantations varied widely. In most descriptions, they were approached as workers, but female authors in particular tend, often based on family histories, to portray women as victims who transformed into heroines (the so-called 'strong' women), while acknowledging their agency (Hoefte 1998; Lalmohamed 1992).

However, the recent picture of life on the plantations shows that the women were very different: they ranged from what from a Western perspective could be called 'docile' women to highly recalcitrant women who put their own individual interests first and were not particularly bothered by the accepted norms of decency (Kumar 2017; cf. Fokken 2018, pp. 170–174). In many cases, they were indeed victims, not so much of the plantation regime but of British Indian men, plantation authorities and other men. Women were not always defenceless as some men stood by them (sometimes even uprisings broke out), while in other cases they could stand on their own.

The scarcity of women not only led to great freedom and agency among women but also to insecurity among men. This often resulted in the use of violence, murder or manslaughter and suicide. In British Guiana, the phenomenon of wife murdering took on epidemic proportions—so much so that the government developed policies to promote family formation (Mohapatra 1995). According to Comins (1892, p. 25), who speaks of 'wife chopping', this phenomenon was incidental in Suriname. Violence against women constituted the male mirror image of *agency*. It consisted not only of jealousy but also the sexual exploitation of women. And here again, the question arises about the scale on which violence and exploitation occurred and the historical significance of women.

The caste system could not be maintained because of the plantation regime that treated everyone as equal, because of the lack of women coming from the same caste, because regulations on purity could not be complied with and because differences between Hindus and Muslims and among currents within Hinduism slowly lost their sharpness. These circumstances also strengthened the position of women, who were able to escape the restrictive provisions and gain a more equal position. In addition to this, there were two other factors that influenced their position favourably, namely the protection of men on the plantation and the possibility of settling outside the plantation once the contract had ended. The equalisers were thus conducive to the amalgamation of the different groups of workers.

4.4 Social Life

Very little has been written about social life on the plantations. Most studies refer to developments after 1920, i.e. the period after immigration. By then, the bulk of the British Indian stayers had settled outside the plantations. Authors of an older generation such as Karsten (1930) and De Klerk (1953) assume, as do some contemporary authors (Bhagwanbali 2010; Choenni 2016), that Hindostani life as known in the districts after World War II was a continuation of life on the plantations. For instance, Bhagwanbali (2010, p. 107ff) paints a romantic picture of workers' leisure activities on the plantations where they sang songs, entertained each other by sharing riddles and telling stories, through games and sporting activities, and by celebrating festivities that could sometimes degenerate into fights. Even though these events took place on the plantations, their significance should not be overestimated: the health of many workers was poor, the number of people was small and they preferred to spend the little free time they had growing their own crops or other things. The previous chapter showed that life on the plantations was characterised by misery and sorrow rather than cheerfulness and leisure activities.

Hassankhan (2016) reports that there were few, if any, Hindu and Muslim places of worship on the plantations. Nevertheless, numerous religious rituals were performed. Authors cite, besides *phagwa* and *Diwáli* among Hindus, *Eid-ul-Fitre*, *Eid-ul-Ada*, *taziyá* and *Munarram* among Muslims as the most common religious festivals (Choenni 2016; Fokken 2018; Hassankhan 2016). In addition, there were religious rituals that were unlinked to holidays such as a *kathá* or a *kitáb* because those religious gatherings were individual initiatives. 'During the period of indenture, Hindus and Muslims lived together and participated in each other's festival and functions ... There were many intermarriages between Hindus and Muslims because of the shortage of women ...' (Hassankhan 2016, p. 195). This social and religious amalgamation could also explain some of the work absenteeism: Muslims not working on Hindu holidays because they participated in the festivity and vice versa.

A unifying element was the food. Part of the food was obtained from the plantation shops, which were mostly run by Portuguese and Chinese. The large plantations, such as Mariënburg, had their own plantation shops. Hoefte (1998) believes that food prices were exorbitantly high to prevent the wages paid from leaving the plantation. Why that would be of interest to planters is unclear. More important is the claim that workers started growing their own food because food prices were so high (Bhagwanbali 2010; Fokken 2018). However, such substitution seems unlikely. The shops on the Mariënburg and Zoelen plantations, for example, sold bread, butter, flour, cheese, potatoes, rice, peas, beans, bacon, canned sardines and the like in addition to beer, gin and rum (Hoefte 1998, p. 156)—all imported products that could not be substituted in the microeconomy of the plantation.

The production of other goods such as rice, coconut oil, ghee, sugar, eggs and vegetables was in the workers' own hands. This production was a tradition rather than a substitution of expensive retail products. My interpretation is supported by Fokken (2018, p. 178), who writes: 'The immigrants used their cooking skills to make dishes that matched their preferences to what was available. They explored new culinary spectra, while generally staying close to ingredients and modes of food preparation they were familiar with. Their use of spices and certain ingredients, such as pulses, set them apart from Afro-Surinamese, Chinese or Javanese plantation residents.' The set of ingredients, the dishes and the occasions at which they were served promoted the same lifestyle that made the cultural bond between the migrants self-evident.

Social bonding was also promoted by visiting. However, workers needed a pass to leave the plantations, except on days off or when they were not assigned a task. It is unknown when the pass system was abolished. It is likely that the system gradually fell into disuse. Around 1915, i.e. towards the end of the indentured system, freedom of movement increased. Eyewitnesses reported that '[a]ll facilities are given for meeting friends, but the absence of roads in most country districts is an obstacle to free social intercourse. Waterways are the ordinary means of communication and, though estates make no difficulty about lending small boats and Indian settlers will place boats at the disposal of friends, obviously intercourse with Indians on other estates or settlements is much more restricted than in places where a walk of one to three miles brings a labourer to another estate or to an Indian community... Visits are also paid to other plantations or settlements. We receive no complaints about restrictions on free movement' (McNeill and Lal 1915, p. 16).

Social life was also defined by mutual services that promoted bonding. A typical example is shown in Photo 4.1, but similar activities must have been extended on a larger scale. Workers were familiar with professions from British India such as those of barber, carpenter, goldsmith, tailor, potter, corpse washer, priest, cook (a special expertise at parties), practitioners of white and black magic and the like. These services and activities are mentioned en passant by eyewitnesses (i.e. Comins 1892).

In Chap. 2, I pointed out the various artisanal occupations, whether or not related to the caste system, that the workers performed in their agrarian milieu (see also De Klerk 1953). These skills, which also belonged to the human capital of the workers, although non-agrarian, were employed on the plantations. It is unknown whether this was a commercial or a community service. It is likely that there were quid pro quos in kind in return for service, an exchange that should have increased the internal bonding.

Social life on the plantations also depended heavily on the time available, specifically at weekends and on public holidays, when people wanted to attend *kathás*, *kitábs*, marriages and religious gatherings. Free time was not always spent on social activities, however, since workers also practised agriculture.

With regard to weekends, Comins (1892, p. 26) writes: ‘On Saturday afternoon labourers knock off work at 1 P.M., and the pay is distributed, cutlasses and hoes sharpened, &c. On Sunday one or two hours are employed in any cleaning of the premises necessary, but outside the mills the only work done is the tending of stock

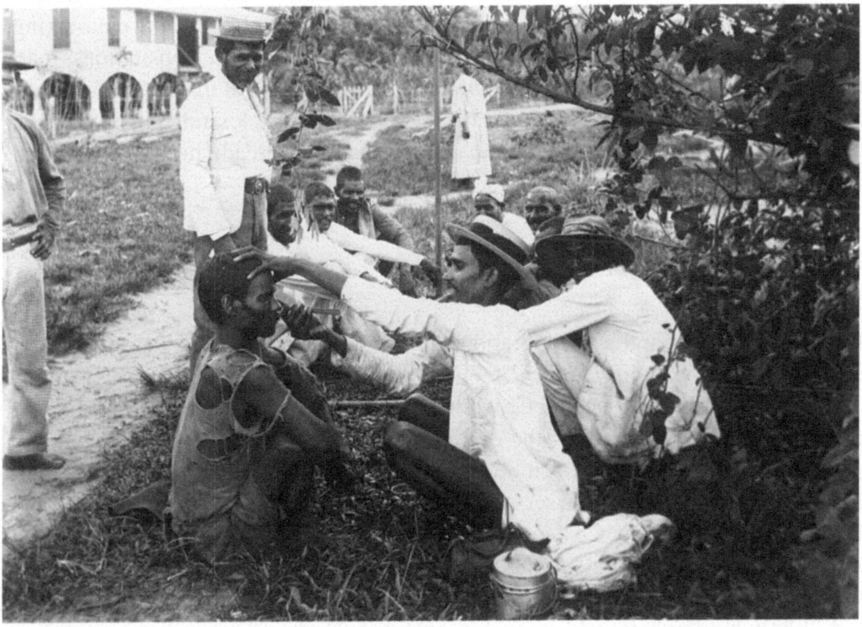

Photo 4.1 Coolies shaving each other, ca. 1920

and the rowing of the employer's boat. Some take more rum than is good for them on Saturday or Sunday, a habit they have acquired from the blacks, and consequently not much work is done Monday morning.' This utterance should not be seen as a description of set habits. Over three decades later, i.e. after the end of the contract period, Lampe (1927, p. 413) wrote: 'The British Indians sometimes use gánjá (cannabis sativa) as a stimulant, the importation of which has been banned since March 1918. The abuse of starken liquor is generally not very great.'

In Surinamese circles, mention is often made of the emergence of a community language from the many local languages from British India called 'Sarnami Hindostani' (Gautam 1999; Gobardhan 2001; Marhé 1985). It is plausible to assume that as with social and religious amalgamation, linguistic unification must also have begun in the depots and on the ships and plantations. Everyday intercourse, marriages, participation in each other's religious and other festivities, such as the relationships entered into as *dipuá-bhái* and *jaháji-bhái* or *bahinin* and the gradually growing numerical preponderance of the British Indian population on the plantations, probably contributed to the development of a Surinamese-Hindostani language.

The ethnic community on the plantations was thus nurtured by family formation, religious expressions, food, the freedom to visit each other, mutual service and the development of a language. It was certainly not a complete community at that stage because institutionalisation was limited, but it was an ethnic community in the making, an embryonic community. These developments continued after the plantation period, or at least, there are no reasons to believe that they were aborted. Karsten (1930) gives a brief description of British Indian life in the colony, focusing on pandits, religion and language, the condition of women, food, drink and oath-taking, home furnishings, national dress, hairstyle, family sense, parties, visitations, manners, marriage, birth, burial and association life. It is doubtful that these mores and customs existed on the plantations, but his description does reflect the fact that around 1930 numerous internal ethnic institutions were already highly crystallised.

4.5 Expressions of Emotion

Despite the long pedigree of migration studies, little attention has been paid to the emotional aspects of migrant life. Emotions mark group culture: some ethnic groups cherish melancholy and nostalgia, while others are more characterised by fun and entertainment. In the study of the diaspora of contract workers, some attention is slowly beginning to be paid to emotional traits such as sadness, homesickness, loneliness and nostalgia, but these serve mainly to illustrate the hard life of the migrants, to portray a condition of existence of the first migrant workers. Typical of this approach are some prominent works on the development of an ethnic group such as those by Hassankhan (2013), Seecharan (2011), Fokken (2018), Klass (1961) and Lal (2000), but the list could easily be expanded. The precise significance of these emotions for ethnic group development has hardly been reflected upon. A few emotions recur frequently in reflections on British Indian labourers, however.

The first is that the migrants were misled by recruiters, causing them to leave their kith and kin and even to abandon them. It is not clear exactly what the emotion is here: the anger evoked by the deception or the sadness of the migrants because they would not see their loved ones again. The deception should have come to an end after 5 years, although the workers had the right to demand immediate return when they arrived in Suriname (Choenni 2016).

However, little use was made of this option. Even more curiously, the workers were able to return after the contract had expired. The permanent settlement in Suriname, resulting in the workers not seeing their loved ones again, is thus difficult to explain as an effect of deception by recruiters. So this emotion must have developed and spread later on among Hindostanis and should be seen as a 'heritage'' of their oral culture. A second recurring type of emotion is grief over the loss of those left behind in British India. Most historical studies have focused on circular migration, remittances and their spending.

In recent years, interest has increased in the emotions of those left behind. The title of the book edited by Majumder (2010), *Kahe Gaile Bides* (*Why did you go abroad?*), is telling. Majumder (2010) especially emphasises suffering and sorrow. The book contains many lyrics of songs in which abandoned women mourn the absence of male relatives (husband, father, brother or son). The author argues that this culture of grief developed in parallel in British India and Suriname. Such emotions were recorded early in history in the form of songs (Grierson 1883). After that, things were quiet for a long time. It was not until 1983 that Brij Lal raised songs again as a source of emotions and a perspective on the past. Since then, songs have featured in almost every historical consideration and have been interpreted as expressions of emotion, especially of those left behind (Choenni 2016; Hassankhan 2013; Majumder 2010; Narayan 2017; Roopnarine 2007; Singh 2016). Some of these authors tend to assign explanatory meaning to the lyrics of the songs.

A third type of emotion is more cheerful and reflects the celebration of ritual feasts in which singing plays a central role. These would include songs sung during sowing, planting and harvesting, drama and musical performances belonging to all kinds of festive events and songs marking the changing of seasons (for an overview, see Arya 1968). Majumder (2010, p. 43) described this 'festive culture' as follows: 'At the time when people from these regions (Uttar Pradesh and Bihar, RG) were recruited as indentured labourers, they sang these songs on their way to Surinamese society. The oral narratives of these stories by the migrants helped them to settle down in Surinamese society. Out of these, many folk cultures and traditions of a variegated nature are still alive.' This is a rather upbeat account of the recruitment and stay in Suriname.

But although the theme, the emotions of those involved in migration, is of great importance, the sources have rarely been critically scrutinised. The emotions of those left behind are derived from observations and interpretations of colonial officials in the former British India such as the much-cited report by Grierson (1883), who recorded the views of various social categories in the emigration area. Some authors dismiss the observations of British colonials because they belonged to the rulers. That is Fokken's reason (2018) for reading colonial data against the grain. Such a position tends to ignore the fact that clashing views also existed among colonial rulers and most likely among the recruits too.

Quoting from colonial reports to one's heart's content is therefore not convincing; the quotes require more support or contextualisation. Using individual writings is also a dangerous hobby. In Surinamese historiography, the uncritical use of the diary of former indentured labourer Munsi Rahman Khan (2005), one of the few migrants who could write, is common (Choenni 2016; Fokken 2018). The question here is whether Khan's observations were accurate, supported by other sources, whether his emotions played a role in the writing down of the texts and in what context he wrote the book.

Songs in which women in particular lament the departure of their spouses are also discussed as emigration emotions (Hassankhan 2013; Lal 2000; Majumder 2010). These songs would still be sung in Uttar Pradesh and Bihar in 2017 (Narayan 2017). Such an observation does not clarify exactly who sang the songs: the relatives left behind or members of later generations. The exact origin of the songs, i.e. the source, is a problem. Moreover, it is questionable whether songs are a direct expression of grief and loss. Love, grief and longing are often sung, not so much because the singers are weighed down by these emotions but because they are an (artistic) expression. By this I do not mean to say that emotions cannot be expressed in song and music. But to assume without question that lyrics reflect a historical reality of Uttar Pradesh and Bihar is problematic because these utterances may be the product of later generations. In that case, they represent the production of oral culture, a heritage.

This heritage is of a very uneven quality. It mostly comprises family histories or stories describing themes such as deception of the migrants by the recruiters, the fate that befell them, the feeling of being lost in the foreign country and, increasingly, the heroic resurrection from the mire of the plantations and the countryside. The authors of these stories are often direct descendants of the immigrants who write about their family history rather than a collective experience as suggested in the song lyrics. Most lyrics in the songs, which incidentally can be found across diaspora countries, express emotions other than heroic self-satisfaction: guilt, homesickness, sadness and loneliness (Lal 2000; Majumder 2010; Narayan 2017).

That the Girmitiyas have known suffering and grief need not be denied or trivialised. However, explanatory power cannot be attributed to lyrics of songs or family stories. Loss, grief and regret may have arisen retrospectively and be undisputed as fact without offering an explanation. Moreover, memories may also be distorted. Therefore, it is risky to causally link the emotions living in the group to historical events. My doubt is reinforced by the observation that none of the contemporary singers, biographers or historians belong to the first generation of migrants, a few to the second and by far most to the third and later generations. Nor can it be maintained that a song offers a historical explanation. A song may have been sung for the singing, for the music, for the dancing, for the recognisability or simply for fun. However tempting, not too much should be read from a song.

In the unlikely event that an indentured labourer left his/her kith and kin because of deception or acted in ignorance, the question is: how come they were so well prepared for a long journey? An example is the currently very popular *baithak gáná* music and dance (Elahi 2017). *Baithak gáná* music is played using three instruments: a harmonium, a *dantál* and a *dholak*. How did these instruments come into the possession of the British Indian indentured labourers if they thought they were going away for a short time or were simply deceived? Or did they 'order' them from Suriname later, indicating that they knew they would not return at short notice? The point here is that no historical significance can be attached to the stories of grief and loss, either of earlier or later generations. This body of stories and songs should be counted as part of the heritage of a migrant community, not of scholarly discussion.

Emotional expressions also include the use of alcohol and narcotics. Comins (1892, pp. 25) noted that '... immigrants at an early period of their residence learn to drink spirits, and to excess, from which much crime and disorder result and the loss of the immigrant's savings'. He suggested that the indentured labourers learned to drink in Suriname, suggesting that the British Indians were practically unfamiliar with the use of alcohol. Comins also opined that the use of narcotics was infrequent. Almost 40 years later, government physician Lampe (1927) would report the same thing: both drinking and the use of narcotics among Hindostanis was moderate. In contrast, De Klerk (1953, p. 146ff) argued that the use of alcohol and narcotics was notable and showed little improvement over time.

According to Bhagwanbali (2010), aggression (by men among themselves, against *sardárs* and plantation authorities, and against women) would be due to the high intensity of frustration. These manifestations may have had their causes partly in the labour situation, but it is doubtful that every abuse of a woman or every conflict between men can be blamed on working conditions. In the same manner, suicide can be put into perspective: not all suicides are a result of past losses. Whatever the reasons, life in remote plantation colonies came at a price that was independent of earnings, such as health, plantation regime, finding a suitable partner and the like. The labourers could buy some services and mitigate some effects, but the price paid was related to their stay in the colony rather than their luck or suffering (Choenni 2016).

Fokken (2018, p. 205ff) reports that the maximum number of letters sent was 1244 (in 1910) and the number of letters received was 1011 (in 1908). The numbers fluctuated over the years and depended partly on the arrival of new immigrants and re-indenture. But the content revealed few emotions. By far the most common message in the letters was a request to return, to give a sign of life or for financial support (see also Choenni 2016, pp. 573–575). Fokken also notes that sending letters was expensive compared to incomes at the time, especially considering that the lion's share of the migrants, like their relatives in British India, could neither read nor write. So these services had to be paid for. What is striking is the discrepancy between the functionality of the letter contents and the emotion in the later songs and stories. This observation also supports the interpretation that while grief was present, it was cultivated by later generations.

4.6 Interracial Relations

During the formative phase of the British Indian ethnic community, social relations with other ethnic groups were important. It is unlikely that British Indians and Creoles lived together on a plantation for 5 years and did not adopt cultural elements from each other. But there is no evidence that such adoption occurred on any significant scale. 'Interracial marriage was a rarity, averaging about two a year in the period 1892–1917, with a high of five in 1909. About three-quarters of these marriages involved the union of people from the British Indian and Creole groups. The remaining marriages were contracted between Javanese and Chinese, and Javanese and Creoles. No Javanese-British Indian unions were recorded during this period' (Hoefte 1998, p. 178). This observation is no more than an indication, because the absence of formal unions does not exclude the possibility that informal relationships did exist.

The low mutual influence requires an explanation. A first impetus for this is the gradual decline of the Creole population on the plantations (see Annex 4.2). The overview presented in the annex shows that the number of Creole workers on the plantations was already declining before the immigration of British Indian immigrants, while that of immigrants was increasing. In the first year of British Indian immigration (1873), the number of immigrants on the plantations doubled. By the end of the nineteenth century the number of Creole workers had gradually decreased to less than half. The increase in the total number of workers (see last column of Annex 4.2) in the last decade was due to an economic revival and concomitant demand for British Indians and, after 1890, Javanese immigrants. Creole workers gradually left for balata mining, the gold industry and railway construction (Willemsen 1980, p. VIII). This meant that contact between the ethnic groups became less and less frequent.

It should also be noted that ethnic group formation worked against any blending. Strong ethnicising forces existed on the plantations, such as the caste consciousness, eating habits, religious and other rituals, language and social relations. These elements became more firmly established over time. The continued orientation towards British India also contributed to the internal cohesion of the Hindostani ethnic community (Gowricharn 2013). Despite the proximity of relatively large numbers of members of other ethnic groups, blending was not natural under the aforementioned circumstances. In addition, De Klerk (1953, pp. 210–211) pointed out that the British Indians rejected racial mixing (with Creoles, RG) early on in their history on the grounds that by mixing, 'their popular power will not (will) be levelled in their favour and thereby the Surinamese commune as a whole, as what it would gain in unity, it would lose in performance'.

4.7 Conclusions

This chapter has addressed the emergence of an embryonic community formation on the plantations. The starting point was latent ethnicity as manifested in several similarities and relationships that emerged during the stay in the depots in British India, the journey to Suriname and the stay on the plantations. While specific terms existed for the relationships in the depots and the journey (*dipuá* and *jaháji*, respectively), no specific designation exists for the relationships that developed on the plantations. Most relations fostering the embryonic community formation presumably took place in the proletarian phase, covering both the recruitment and transportation of the workers and their society on the plantations.

Most important was the great equalising influence of the plantation system: everyone was basically a worker and treated as such. The system of indentured labour was the 'great leveller of hierarchy' that seriously damaged the caste system and enabled women to enter into more equal relationships. The small population size was also instrumental in breaking down traditional hierarchies. Many customs surrounding the caste system could no longer be maintained on the plantations due to the small size of the Girmitiya population, while some women saw their bargaining power improve. The arrival of new workers did not significantly increase the population on the plantations because many workers left or settled outside them. The combination of the small population, a limited number of women and the equalising labour regime was decisive for the emergence of the Surinamese-Hindostani community.

The breakdown of traditional institutions did not mean that they disappeared altogether. On the plantations, new institutions were established or old ones continued in a modified form. This was true for the performance of rituals that required members of higher castes without reproducing the caste system, for the celebration of numerous religious and community festivals, for the emergence of Sarnami Hindostani, for the provision of mutual services and individual help—all things that reflected a modest social life on the margins of the plantation. Although the perspective of Hindus dominates in the portrayal (which can be explained by the numerical preponderance of this religious community), Muslims were closely involved in the development of ethnic community formation, especially through co-habitation.

There are no indications that cultural elements from other ethnic groups were included in this institutionalisation process. The institutionalisation of customs, adapted or otherwise, is a normal occurrence for migrant workers. Specifically, Girmitiyas were weighed down by feelings of homesickness, sadness and nostalgia. Later generations used these emotions to produce songs about various aspects of (imagined) plantation life. This would have given their ethnicity a specific colour—a characteristic also reflected in music, song, drama and storytelling. Although these expressions boomed from the second half of the twentieth century onwards, it is doubtful whether they represent the first generation's life accurately. Rather, it is likely that traditions, nostalgia and sadness have been cultivated by subsequent generations and commercialised as artistic expressions. If this doubt makes sense, then this body of utterances belong to a created Hindostani heritage and cannot be assigned explanatory significance. They are thus an indispensable part of Hindostani community formation.

References

Alexander C, Chatterji J, Jalais A (2018) The Bengal diaspora. Rethinking Muslim migration. Routledge, London

Arya U (1968) Ritual songs and folksongs of the Hindus of Surinam. Doctoral dissertation, University of Utrecht

Bhagwanbali R (2010) Hindoestaanse migranten onder het indentured labour systeem naar Suriname, 1873–1916. De nieuwe awatar van slavernij. Amrit, Den Haag

Carrier J (2018) Moral economy. 'What's in a name? Anthropological Theory 18(1):18–35

Choenni C (2016) Hindostaanse contractarbeiders 1873–1920. Van India naar de plantages in Suriname. LM Publishers, Volendam

Comins DWD (1892) Note on emigration from the East Indies to Surinam or Dutch Guyana. Printed at the Bengal Secretariat Press, Calcutta

De Klerk CJM (1953) De Immigratie der Hindostanen. Urbi et Orbi, Amsterdam

Elahi J (2017) Hindostaanse gemeenschapsdansen. In: Gowricharn RS (ed) Onbeschreven Erfgoed. Perspectieven op de Surinaams–Hindostaanse Cultuur. LM Publishers, Volendam

Emmer P (1984) Een goede keus? De migratie van vrouwelijke contractarbeiders van Brits-Indië naar Suriname. OSO, Tijdschrift voor Surinaamse Taalkunde, Letterkunde en Geschiedenis 3(2):213–234

Emmer P (1986) The meek Hindu: The recruitment of indentured labourers for service overseas 1870–1916. In: Emmer P (ed) Colonialism and Migration: Indentured Labourers Before and After Slavery. Martinus Nijhoff, Dordrecht, pp 187–207

Emmer P (1987) The position of Indian women in Suriname. Boletin de Estudios Latino americanos y del Caribe 43:115–120

Fenton S (2003) Ethnicity. Key concepts. Polity Press, Malden

Fokken M (2018) Beyond being koelies and kantráki. Constructing Hindostani identities in Suriname in the era of indenture 1873–1921. Doctoral dissertation, University of Groningen, Groningen

Gautam M (1999) The construction of the Indian image in Surinam: Deconstructing colonial derogatory notions and reconstructing Indian identity. In: Mahin G, Narine D (eds) Sojourners to Settlers: Indian Migrants in the Caribbean and the Americas. Windsor Press, New York, pp 125–181

Geertz C (1983/2000) Local knowledge: Further essays in interpretative anthropology. Basic Books, New York

Gobardhan L (2001) Onderwijs Als Sleutel Tot Maatschappelijke Vooruitgang. Een Taal- En Onderwijsgeschiedenis Van Suriname, 1651–1975. Doctoral dissertation, University of Leiden

Gowricharn RS (2013) Ethnogenesis: The case of British Indians in the Caribbean. Comp Stud Soc Hist 55(2):388–418

Grierson GA (1883) Report on colonial emigration from the Bengal presidency. Government of Bengal, Bengal

Hassankhan M (2013) Kahe Gaile Bides – Why did you go overseas? An introduction in emotional aspects of migration history: A diaspora perspective. In: Mahabir K (ed) Caribbean issues in Indian diaspora. Serial Publications, New Delhi, pp 3–35

Hassankhan M (2016) Islam and Indian Muslims in Suriname: A struggle for survival. In: Goolam V, Roopnarine L (eds) Indentured Muslims in the diaspora. Identity and belonging of minority groups in plural societies. Manohar, New Delhi, pp 183–228

Hoefte R (1987a) Female indentured labor in Suriname: For better or for worse? Boletín de Estudios Latinoamericanos y del Caribe 42:55–70
Hoefte R (1987b) The position of female British Indian and Javanese contract laborers in Suriname. A last word. Boletín de Estudios Latinoamericanos y del Caribe 43:121–123
Hoefte R (1998) In place of slavery. A social history of British Indian and Javanese laborers in Suriname. University Press of Florida, Gainesville
Jayawardena C (1963) Conflict and solidarity in a Guyanese plantation. The Athlone Press, London
Karsten R (1930) De Brits–Indiërs in Suriname. Een korte schets benevens een handleiding voor de beginselen van het Hindi. Martinus Nijhoff
Khan MR (2005) An autobiography of an Indian indentured labourer. Amrit, Den Haag
Klass M (1961) East Indians in Trinidad. A study of cultural persistence. Waveland Press, Long Grove
Kumar A (2016) Feeding the Girmitiya 1: Food and drink on indentured ships to the sugar colonies. Gastronomica: The Journal of Critical Food Studies 16(1):40–51
Kumar A (2017) Coolies of the Empire. Indentured Indians in the sugar colonies 1830–1920. Cambridge University Press, Cambridge
Lal B (2000) Chalo Jahaji. On a journey through indenture in Fiji. Australian National University Press, Canberra
Lal B (2006) Fiji. In: Lal B, Reeves P, Rai R (eds) The Encyclopedia of the Indian diaspora. Singapore, Editions Didier Millet, pp 370–382
Lalmohamed B (1992) Hindostaanse vrouwen: de geschiedenis van zes generaties. Van Arkel, Utrecht
Lampe PHJ (1927) Suriname. Sociaal–Hygiënische Beschouwingen. De Bussy, Amsterdam
Majumder M (ed.) (2010) Kahe Gaile Bides (Why did you go overseas?). In: On Bhojpuri migration since the 1879's and contemporary culture in Uttar Pradesh and Bihar, Suriname and the Netherlands. Mango
Marhé RM (1985) Sarnami Byäkaran: een elementaire grammatica van het Sarnami. Stichting voor Surinamers voor Den Haag, Den Haag
Mayer A (1966) Quasi-groups in the study of complex societies. In: Banton M (ed) The social anthropology of complex societies. Tavistock Publications, London, pp 97–122
McNeill J, Lal C (1915) Report to the government of India on the conditions of Indian immigrants in four British colonies and Surinam, Part II. London
Mohapatra P (1995) Restoring the family: Wife murders and the making of sexual contract of Indian immigrant labour in British Caribbean colonies 1860–1920. Stud Hist 11(2):227–260
Narayan B (2017) Culture and the emotional economy of migration. Routledge, London
Roopnarine L (2007) Indo-Caribbean indenture. Resistance and accommodation, 1838–1920. University of the West Indies Press, Mona
Samaroo B (2006) The Mahatma in the Caribbean. In: Gowricharn R (ed) Caribbean transnationalism. Migration, pluralization, and social cohesion. Lexington Books, Lanham/Boulder, pp 139–153
Seecharan C (2011) Mother India's shadow over El Dorado: Indo-Guyanese politics and identity 1890s–1930s. Ian Randle Publishers, Kingston en Miami
Singh N (2016) My life my story: Narratives of women left behind. In: Hassankhan M, Roopnarine L, Mahase R (eds) Social and cultural dimensions of Indian indentured labour and its diaspora: past and present. New Delhi, Manohar, pp 25–50
Snellen E (1933) De Aanvoer van arbeiders voor den landbouw in Suriname. Doctoral dissertation, University of Wageningen
Willemsen G (1980) Koloniale politiek en transformatieprocessen in een plantage-economie. Suriname 1873–1940. Doctoral dissertation, Erasmus University of Rotterdam

Part II
From Labourer to Peasant

Chapter 5
The Rise of the Peasantry, 1880–1920

Abstract This chapter deals with the rise of a Girmitiya peasantry. That was fostered by the decline of the plantation sector, which made workers redundant and land available. The colonial government embarked on a colonisation policy aimed at increasing the population to create an ample labour supply that could push down wage levels. The settlement was made attractive to British Indian immigrants by giving them a piece of land and a premium of *f100* if they waived their free return to India. The argument is structured as follows. Section 5.2 discusses the deterioration of plantation agriculture, while Sect. 5.3 discusses the shift from labour politics to population politics, mainly consisting of an increasing settlement of Girmitiya peasants. Section 5.4 builds on this development: the Girmitiya settlers became Suriname's main food suppliers, especially rice. Section 5.5 zooms in on the fragmented agricultural landscape in the first three decades of the twentieth century. During this period, other economic sectors were practically depleted and hopes were pinned on rice exports. Section 5.6 discusses the problems of rice production for export and policies related to it.

Keywords Settlement · Wage policy · Girmitiya peasants · Rice producers · Agricultural policy

5.1 Introduction

In most colonies, plantations remained in operation well into the twentieth century, so the need for workers persisted. Land was monopolised by the plantation owners or was made available to the workers in small amounts. At best, this led to two segments among immigrants, one that continued to work on the plantations and one that settled as peasants (Allen 1999; Mandle 1973; Mayer 1973). Peasants often grew the same product as the plantations or sold it to the planters and other buyers. Some peasants even grew into small planters who employed immigrants, as in Mauritius (Allen 1999). In Suriname, Creole peasants developed into commercial

R. Gowricharn, *The Girmitiya Peasants in Suriname*,
https://doi.org/10.1007/978-3-031-67961-2_5

middle-class farmers, focused mainly on coffee and cocoa cultivation (Heilbron 1982).

This dependency has been best articulated in the Caribbean by authors such as Beckford (1972) and Mintz (1983). In their view, peasants were an appendage of plantations; the work as a peasant was an afterthought, often a strategy of planters to secure labour by making land available. In this perspective, smallholders are not autonomous agricultural producers but workers in the service of the plantation (Goodman and Redclift 1981). In Suriname, however, a complete decoupling of plantations and contract workers originated. As a result, the peasantry became highly autonomous. The question then becomes how the peasantry in Suriname developed separately from the plantations.

This chapter deals with the policy pursued in the period 1895–1930, the period in which the preconditions arose under which the peasantisation took place. The main precondition consisted of the decline of the plantation sector, which resulted in plantations closing down, fewer workers being needed and land being set free. A second precondition consisted of a colonisation policy aimed at increasing the population to create an ample labour supply that could push down wage levels. To this end, settlement was made attractive to British Indian immigrants by giving them a piece of land and a premium of *f100* if they waived their free return to India. Since all peasants cultivated different products, the smallholder sector soon harboured several categories of ethnic peasants (Dutch, British Indian, Creole and Javanese peasants), each utilising their own agricultural human capital. Thus, they were not in a competitive relationship with each other.

The next section discusses the deterioration of plantation agriculture. This development reveals how, despite an initial increase in the old export products, plantations declined in importance. Section 5.3 discusses the shift from labour politics to population politics. This consisted mainly in increasing the population by encouraging the settlement as peasants that occurred under specific conditions and enabled the settlement of all Girmitiyas. Section 5.4 builds on this development: the Girmitiya settlers became Suriname's main food suppliers, especially rice. Section 5.5 zooms in on the fragmented agricultural landscape in the first three decades of the twentieth century. During this period, other economic sectors were practically depleted and hopes were pinned on rice exports. Section 5.6 discusses the problems of rice production for export and policies related to it. The conclusion assesses the development of the Girmitiya peasantry.

5.2 The Decline of Plantation Agriculture

In most plantation economies, capital, technology and labour came from outside to drive large agricultural enterprises. Initially, local labour was brought in whenever possible. In the Caribbean, this consisted of native Indians. However, they barely survived the plantation regime. Several authors point out that in a second phase, white young men were contracted as labourers who were employed as bonded labour for meagre wages and free board and lodging, especially in England, the US and France, and some colonies (Ballagh 2004; Northrup 1995; Reddy 2016). More famously, instead of recruiting white contract workers, planters brought enslaved Africans to the Caribbean for labour on plantations. In 1808, England banned the slave trade under pressure from the abolitionist movement. Slavery was abolished in its colonies in 1833, in the French colonies in 1848 and in the Dutch colonies in 1863.

Plantations mainly produced sugar, as well as coffee, cocoa, cotton, tobacco and timber for the world market. The only domestic factor of production, the land, had to be productive and brought under cultivation (think deforestation, tillage, construction of canals and roads) to be of any value. De Cock Buning (1926, p. 52) reported that Suriname had about 600 plantations at its peak, between 1790 and 1800. Van Stipriaan (1993, p. 33) arrived at a lower number of plantations of over 400, but he limited himself to sugar, coffee and cotton plantations in the export sector. Many of the plantations were owned by administrators, some of whom were entrusted with the care of 50–60 plantations. Most planters were Dutch, but English, Scottish, French and Germans were also well represented among them.

For various reasons, the number of plantations in Suriname decreased. The most important of these was the crisis on the Amsterdam stock exchange in 1773, which caused many Surinamese plantations to go bankrupt. Maroon attacks also affected the profitability of many plantations, but it is not clear whether there was a link between the maroon attacks on the plantations and the stock market crisis (Hira 1982). A new wave of closures took place in the decades preceding the abolition of slavery in Suriname. These were prompted not only by economic weakness (debts, mortgages and mismanagement) but also by disbelief that the plantation colony could be saved without slaves. This led to a departure of plantation owners, who sold their plantations or left their management to administrators and supervisors.

According to van Lier (1977), this flight of plantation capital as well as better-motivated entrepreneurs were major propellers in the decline of the plantation sector. In addition to all these propellers, Panday (1959) attributes the closing off of the European market by Napoleon, who wanted to protect the rise of beet sugar from the competing cane sugar, as one of the major reasons for the decline of the slave economy. Hence, according to the prevailing story, slavery was abolished because the plantations were no longer profitable (van Lier 1977; Willemsen 1980).

This explanation ignores the strong pressure exerted by abolitionist movements in the UK and the US to abolish slavery and possibly replace slaves with other forms of labour (Northrup 1995; Reddy 2016; Siwpersad 1979). Sometimes the declining share of sugarcane plantations in the total world sugar production is cited as a reason to abolish slavery. In 1839, the percentage of cane in the total world sugar production was 95.1 and 30 years later it had fallen to 68 (Deerr 1950, p. 490). Annex 5.1 shows Suriname's main exports over a century, from 1760 to 1860. Since production fluctuated widely, I have taken the average of 1770–1795 as the base index year (set to 100) to show how exports evolved in the first half of the nineteenth century. The annex shows that sugar exports grew steadily and more than doubled, while cotton exports began to rise around 1816 and peaked around 1825 before falling to their base level in 1850. Apart from 2 years (1859 and 1862), coffee, like cocoa, was below its base level.

So sugar was, and remained, Suriname's main export product for a long time. But its growth lagged far behind that of other sugar producers. In the Dutch colonial empire, Java was the biggest competitor, and in the Caribbean region it was British Guiana. Table 5.1 shows the development of sugar production in these three countries.

According to Table 5.1, sugar production in Java exploded to 117 times its original size in 40 years. In British Guiana, that growth was 1.73 times the production in 1826 and in Suriname 1.15 times as much. Not only are the divergent growth rates striking but the differences in the size of production also stand out. In 1860, the volume of production in Java was more than eight times larger and that in British Guiana was 3.2 times larger than that of Suriname. Judging from the much larger immigration of British Indian workers in British Guiana—238,000 compared to 34,000 in Suriname—this difference can largely be explained by the larger scale of sugar production. The difference in Java was explained by the lower labour costs and higher labour productivity there (Brandon 1925).

In the period up to the abolition of slavery, the number of plantations decreased sharply. Between 1833 and 1863, the number of sugar plantations decreased from 105 to 56; the number of cotton plantations decreased from 31 to 15 in the period 1848 to 1863, while between 1832 and 1857, the number of coffee plantations decreased from 170 to one (Panday 1959, pp. 78–81). The planted area also decreased. In 1848, for example, there were 284 plantations that had cultivated a total of 51,210 ha of land. In 1862 and 1872, the number of plantations was 229 and 112, respectively, and the cultivated area was 39,639 and 28,552 ha (Panday 1959, pp. 82, 126, 127).

According to the data in Annex 5.1, the main export product, sugar, increased in the first decade of the nineteenth century. Sugar exports doubled in the period 1810–1840, while cotton exports experienced an even greater increase in the same period. The amount of land under cultivation thus decreased, while production increased. This can largely be explained by the use of advanced production

Table 5.1 Sugar production in Suriname, Java and British Guiana, in tonnes, 1826–1860

Year	Suriname	Java	British Guiana
1826	11,000	1178	43,378
1830	16,500	6647	59,790
1835	18,900	26,158	57,577
1840	17,200	47,040	35,619
1845	14,900	90,962	34,681
1850	15,600	86,519	32,692
1855	17,400	103,963	48,447
1860	16,700	136,153	54,423
1865	7800	137,893	75,356

Source: Deerr (1950, pp. 203, 212 and 224)

technology, especially the number of steam mills, which increased from 36 to 59 in the three decades before the abolition of slavery (Gowricharn 1985; see also Hira 1982, pp. 75–85). But this innovation did not stop the decline of the plantation colony.

The Dutch government was unwilling to make large investments from the mid-nineteenth century onwards. This was linked to the strong orientation towards the most profitable Dutch colony, the Dutch East Indies (Panday 1959, pp. 111–113). Moreover, most Surinamese plantations were burdened by mortgage debts. The decline of the plantations was partly cushioned by the emergence of gold and rubber production around 1870, which was stimulated by tax concessions, the concession of large tracts of land and a minimum of legal restrictions (Heilbron and Willemsen 1980a, b). Among these investors, government officials, planters, rich peasants and other locally well-to-do people were to be found (Heilbron and Willemsen 1977, p. 78).

The ex-slave population also found alternative employment in gold and balata mining. Despite the declining importance of the plantations, the government continued its policy of importing contract workers. Gradually, the ex-slaves on the plantations were replaced by immigrants. But Girmitiyas also left the plantation sector after serving their contracts to try their luck in other sectors. In response to this sectoral mobility, the governorate banned immigrants from working in the gold and rubber sectors at an early stage (1882) (Heilbron 1982, p. 214). As a result, only ex-slaves were able to switch sectors and plantation agriculture was increasingly populated by British Indians, later supplemented by Javanese.

Annex 5.2 summarises the five-year averages of export value over the period 1866–1935. The annex provides a long-term perspective on Surinamese exports, used here as a reflector of key economic sectors. The trends show that the importance of sugar declined sharply in the period 1880–1885, that is, while the immigration of British Indian workers was in full swing. During the same period, the export value of gold and cocoa increased sharply. In both sectors, an increasing share of production was attributed to small producers (Heilbron 1982, Ch. VI). As small peasants produced a significant proportion of these products, the obvious association between exports and plantation agriculture was broken. Gold, cocoa, rubber and coffee outstripped the export value of sugar in later periods. From 1890 to 1920, the combined value of balata, gold and cocoa even fluctuated between 60% and 70% of the total export value.

The rises in gold mining, balata mining and, later, bauxite mining were accidental 'windfalls', i.e. not a result of policy. After all, government policy until the first two decades of the twentieth century was aimed at restoring plantation agriculture. For this period, Willemsen (1980) and Heilbron (1982) describe an elite, entrenched in administration and plantations, whose thinking and efforts hardly showed any shift. Although there had been an understanding that compensatory economic activities needed to be developed for the lagging plantation agriculture, the focus remained on the plantation sector. This ailing situation resulted in persistent deficits in government finances from the mid-nineteenth century, which were supplemented by subsidies from the Netherlands (Willemsen 1980, pp. 325–326).

Poverty among the free ex-slaves was probably ubiquitous, but it was particularly noticeable in the (main) city of Paramaribo, where they had no means of subsistence. The nature of the problem was such that the government took administrative measures to counteract the 'accumulation of freedmen and other minors in Paramaribo' (Bosz and Albrecht 1954, p. 189). To combat their increasing exodus and to meet the growing urban demand for vegetables, these free ex-slaves were offered a piece of land near the city of Paramaribo. The colonial government provided premiums ranging from 100 to 500 guilders to those who had made productive use of the land by growing rice, bananas and vegetables, and raising horned cattle and sheep. This policy was officially implemented in 1842 and marked the transition from slavery to farming. It became the basis of Girmitiya peasantisation.

5.3 Colonisation Policy

The policy of colonising immigrants has been a hotly and long-discussed issue. Roughly speaking, there were two camps in colonial circles. One camp favoured permanent immigration because it hoped that mutual competition among workers would lower wages in the plantation sector. This camp consisted mainly of planters and was strongly represented in government circles. They also wanted to increase the population because they believed that the former slaves were spinning out of control. Therefore immigrant workers were needed (see discussion in De Klerk 1953, Ch. VI; Heilbron 1982; Hoefte 1998, III). The other camp was in favour of engaging the ex-slave population in the labour process. This camp was convinced that ex-slaves could be educated to work (Beck 1924b; Brandon 1925; Van Praag 1877). The diagnoses and proposed solutions came to the wider public first in the form of speeches and official reports and, from the twentieth century, also in articles published mainly in journals such as *West Indië*, and then *West Indische Gids*, later renamed the *Nieuwe West Indische Gids*, and in *De Economist*.

These two perspectives on the colonial economy, focusing on plantations and the population, respectively, were linked by the demand for labour. However, labour demand comprised several issues. The first was the migration of former slaves to gold and balata mining. This caused a scarcity of labour on the plantations that was met by immigration. But a significant proportion of these immigrants turned back or shirked plantation labour, so the labour shortage remained a concern. Raising wages was not an adequate measure because, according to the views prevailing at the time, ex-slaves had an aversion to plantation labour (van Lier 1977, p. 135).

However, this explanation is unconvincing since substantial numbers of former slaves continued to work on the plantations for a long time or devoted themselves to cocoa and coffee cultivation (Heilbron 1982; Hoefte 1998). Moreover, the absorption of the gold sector remained limited: during the period 1888–1917 (i.e. for three decades), the number of workers in gold mining fluctuated between 2500 and 3000. In contrast, the number of small peasants increased from 6375 to 23,881 (Willemsen

1980, p. 201). These peasants comprised a decreasing proportion of former slaves as more and more immigrants settled as peasants.

A second issue consisted of attempts to secure labour needs through colonisation on the plantations or on land located near them. Workers were thus transformed into peasants who, in the most favourable case, had to provide labour services to the plantations during peak periods (Burger 1928; Van de Houven van Oordt 1900).

But this expectation of planters did not always come true. The small cacao and coffee peasants often grew the same main crop as the plantations, which meant that labour requirements peaked at both locations during the same sowing and harvesting periods. These peasants even grew into middle-class peasants who, by the beginning of the twentieth century, had surpassed the production of the cocoa plantations and accounted for a significant share of exports (Heilbron 1982).

However, cheap and flexible labour remained the inhibiting production factor that, according to the colonial administration, hindered the expansion of the plantation sector. Such labour flexibility was further countered by the agreement with the contract workers, who were paid a fixed minimum wage, even during low-season periods. In the words of De Cock Buning (1926, p. 57), '[i]f they (the planters, RG) employ a number of contract workers tailored to peak periods, then they also have to pay them the minimum wage in the slack time ... This lack of elasticity of labour can only be remedied either by the establishment of free workers in the vicinity of the enterprises, located beyond the power of the entrepreneur, or by colonisation on the plantations'. As workers migrated to other sectors and peasants preferred to work on their own farms during harvest periods, this objective was not met.

Against this backdrop, population augmentation was necessary to secure a cheap and regular labour supply. This was done by colonising workers, a policy that officially started in 1895 but in practice began earlier. Abandoned plantations, mostly with the cacao and coffee plantations on them, were set up as settlement sites. The government subdivided these abandoned plantations into plots ranging in size from 1 to 1.5 hectares. Peasants were supposed to grow their own food. The main incentive for settlement was exemption from land tax for the first 6 years and the free granting of property rights after 2 years of regular cultivation. Burger (1928, p. 427) reports that initially very few British Indians took advantage of it because they lost the right of return when settling as peasants and because the land was poorly developed. In 1894, the government bought up the abandoned Alkmaar plantation, which was set up exclusively for British Indians to settle.

Although the land had been cleared and the drainage and road system put in order, enthusiasm was limited. In 1895, the government stipulated that immigrants would forfeit the right of return only on acceptance of the settlement bonus of f100. This scheme encouraged them to settle as peasants. In 1895, 584 British Indians were owners and tenants of domain or private land, and that number had risen to 5093 by 1913 (Burger 1928, p. 428; De Klerk 1953, p. 165). As a result, the development of British Indian peasants did not make a good start until the early twentieth century. 'Because of this milder provision, hundreds of British Indians settled and the importation originally conceived as labour immigration changed into a colonisation system, in which working on the plantation should be considered a learning experience' (Westra 1921, p. 6).

This idea of the 'plantation as an apprenticeship', in line with Alkmaar's set-up, recurs more often in the literature. For instance, the résumé of the Suriname Study Syndicate (Kasteleyn and Kluvers 1919, p. 290) explicitly states: 'Large-scale agriculture is the only school of learning to prepare immigrants for independent agricultural work, i.e. for permanent settlement.' The Encyclopaedie of the Netherlands West Indies mentioned that the behaviour demanded of Hindostani peasants was satisfactorily: 'They clear their own land, if necessary, and build a simple house there. A less cumbersome and cheaper form of colonisation is inconceivable. It is therefore desirable not to encourage re-contracting. An apprenticeship of five years is ample. The sooner the immigrant settles on his own land, the better' (Encyclopaedie 1917/1981, p. 379). The size of the plot was considered sufficient to provide a living for a family, with the technical means current at the time, for 3 days a week. For the remaining 2 or 3 days, a family head had to work on a neighbouring plantation to supplement their income.

Van de Houven van Oordt (1900, pp. 599–600) distinguished three forms of plots on settlement sites. The first, with plots of 0.8–1.6 hectares in size, was intended to train migrants to practise agriculture, i.e. to raise agricultural capital or to train labourers who had come to Suriname at a young age or needed retraining. Here, the land was fully furnished by the government, such as levelling discharges, roads, bridges, dams and sluices. The second form was for independent peasants, and they had to keep the drains, roads and the like in order themselves. These plots averaged 2 ha in size. The third form was analogous to the second except that ownership could be applied for after 2 years of cultivation. When applying for ownership, the size of the plot could increase to 10 ha. In addition, there were plots scattered all over the country to both immigrants and non-immigrants for rent (Van de Houven van Oordt 1900).

To qualify for a plot, applicants had to show proof of good conduct. Land titles were transferable, but an application had to be submitted to the governor to do so. Permits could be revoked for various reasons, such as if the plot was used for purposes other than farming, horticulture or animal husbandry. Furthermore, small loans were granted and there was the possibility of expanding the acreage (Van de Houven van Oordt 1900). By and large, this was the policy regarding settlement sites.

It was not only the government that offered land to applicants. Many planters also made land available to their (former) employees to reduce the cost of maintenance and/or for the production of staple products (Heilbron 1982, Ch. III and IV). This happened mostly on the cocoa and coffee plantations, but increasingly planters also made plots of land available for rice cultivation. Burger (1928, pp. 434–435) reported: 'The first to start doing so was plantation Vreeland in the year 1909. It gave out land for rice cultivation to British Indians, whose contract had expired. The land was cleared and planted by them, but the harvest was not satisfactory, and they left the land because it was unsuitable for rice cultivation.' And on the next page, we read: 'Slowly, several plantations have moved to allow such settlements (as on plantation Vreeland, RG) ... On very many other plantations, small rice plots have been given to the contract workers ... on the days of planting and harvesting is given freely.' In addition, British Indians, alone or with others, bought up land, sometimes even the size of a plantation, but this category involved small numbers (Choenni 2016; Fokken 2018).

Land acquisition was thus relatively easy in Suriname compared to, for example, countries in the Caribbean where access to land was deliberately restricted (Laurence 1994; Shepherd 1993). In this allocation of land, racial preferences played a role. In colonial circles, especially of economists, agricultural consultants, other concerned officials, missionaries and entrepreneurs (these functions were sometimes united in one person), various racial performance images of ethnic groups were in vogue.

A brief sampling of the images: Creoles were considered unsuitable for agriculture because they suffered from the memory of slavery, even though firm side notes were placed on this (Brandon 1925). Javanese were seen as future agriculturists because they were tidy and managed their plots well, with the caveat that they were not very ambitious (Burger 1928). The most esteemed were British Indians, who were said to be miserly and ambitious, and thus better entrepreneurs (Beck 1924a; Leys 1921; Liems 1926). However persistent these images were among the colonial elite—they were debated well into the twentieth century (see De Vries 1965)—there is no evidence that they influenced land allocation. Unlike what Fokken (2018, p. 211) believes, no facilities were specifically for British Indian peasants: all ethnic groups were able to use the existing facilities.

Arrangements to secure labour supply were undermined by the decline of the plantation sector. That decline freed up land and weakened the relationship between plantations and smallholders. An important effect that follows from this delinking was that peasants were unable to acquire supplementary income—which by design was half the intended family income (2–3 days of off-farm work). Agrarian employment declined and the peasant household had to rely entirely on income from their farming activities. Whether their labour capacity was sufficient to get their income, even if they had enough land, is questionable. But even if they managed to double production, peasants needed to be able to sell it. If they did not manage to produce or sell a surplus, they suffered poverty.

Given the closure of plantations, it is not surprising that the size of the farming sector increased while the population on the plantations decreased. Table 5.2 shows the distribution of the population between agricultural and non-agricultural sectors. From 1880, i.e. shortly after the first cohort of British Indian labourers had served out their contracts, the share of peasants was 8%. That proportion rose slowly from 1888 and jumped after 1897 from 12% to 19%. The population on the plantations remained stationary, with a small decline in the 1880s. By 1910, a small quarter of the population was involved in smallholder agriculture while the share in plantation agriculture fell to less than a fifth.

Table 5.2 Agricultural population as percentage of total population, 1863–1910

Year	Plantation	Peasants	Total agrarian sector	Total population
1863	33	Unknown	33	52,963
1873	28	2[a]	30	51,888
1880	20	8	28	52,856
1888	15	12	27	57,365
1890	15	13	28	58,951
1894	19	13	32	58,866
1897	19	12	31	64,372
1902	21	19	40	62,617
1910	19	24	43	92,142

Source: Compiled from Heilbron (1982, pp. 68, 96, 131)
[a]Encyclopaedia of the Netherlands West Indies (1981, p. 441)

5.4 Small Farming as a Subsector

Plantation agriculture declined and smallholder agriculture increased. This development led to two categories of peasants. The first consisted of the producers of coffee and cocoa destined for the foreign market. They were specialised in commercial products for the international market. The second category consisted of a growing number of producers of food crops for the domestic market. These peasants were no longer producers of food only for their own consumption, but also for other consumers.

Initially, the range of crops grown varied little among ethnic groups. The most commonly mentioned market crops of smallholder agriculture were cocoa, coffee, rice, maize, cassava, sweet potatoes, bananas, plantains, coconuts, beans, melons and fruits. Sales of vegetables, eggs and milk also took place. The sale of meat and fish was sporadically mentioned in the colonial statistics, although most peasants ran mixed farms. Chickens, goats, sheep and cattle were also part of the assortment, but government attention to these was low. Also to be mentioned is the fruit culture, often found in urban yards and plots in the districts. Fruits included citrus, plantains and bananas for export (Liems 1923). Other promising fruits were pine apple, orange and soursop. In the first quarter of the twentieth century, agronomists increasingly looked at smallholder farming as a source for exports rather than as a living for peasants. Central in these deliberations were considerations of price and technology.

Significantly less frequently mentioned were cattle farming and the cultivation of small livestock. Leys (1917) argued that cattle farming suffered from inadequate discharge and a lack of supplementary feed. 'Almost half of Suriname's pastures lack satisfactory discharge. We have already pointed out that such pastures are transformed into swamps in the rainy season' (Leys 1917, p. 106). As regards supplementary feed, he noted that British Indian peasants, with their families, spent a lot of time cutting grass: 'Often here one sees only a few cows, kept almost constantly in stalls or tied to a rope, grazing the grass verges of bar tacks and roads.'

Photo 5.1 Hut of first British Indian peasants outside settlement schemes

Seven years later, Leys (1925a) noted that cattle breeding among British Indians had progressed since 1914. Since it was difficult for British Indians who had served out their labour contracts due to World War I to go back, they 'stuck around' in Suriname by settling as peasants. High food prices and reports that things were going badly in British India shaped the decision to stay in Suriname (Photo 5.1).

With the rise in the number of British Indian peasants, the number of cattle increased. 'Every British Indian aspires to have one or more cows in the barn. I would almost say to have in the house. The extremely primitive houses, or rather huts, of the Asians are often arranged in such a way that the cowshed is a part of the house. In the fullest sense of the word, the cow is a pet for the British Indian, for the Hindu it is even a sacred animal, and although this is not the case for the Mahomedans, he still reveres his cow and cares for it better than his wife and children' (Leys 1925a, p. 406). The increased number of cattle made Suriname independent of slaughter cattle imports and even created a surplus of milk. The cattle additionally served to plough and harrow and transport, while they also provided manure. They had a function in the agricultural process, an element that did not receive much attention from agronomists of the time.

In a later publication, Leys (1925b) elaborated further on livestock policy. Among small livestock, he included pigs, goats, sheep, donkeys, horses and poultry (especially chickens and ducks). The entire range of small livestock was characterised by one-sided feeding and inbreeding that caused the breed to decline in fertility, slaughter weight and resistance to disease. Cultivation was not systematic; in fact, according to Leys, these animals were kept without being farmed. The numbers per peasant were small and sales uncertain. What Leys writes for goat farming applies to cattle farming as a whole: 'Goats thrive here. So here and there one or two are seen with small peasants, while some British Indians keep as many as six or eight. There is really no question of breeding goats. Family breeding and far too early mating of goats is the order of the day. The goats are small and weak; the milk yield is insignificant' (Leys 1925b, p. 448).

Given this state of affairs, the government took measures to improve the breeds. It established a farm on the former plantation Beekhuizen where some studbooks were stationed. A studbook buck was stationed at the Cultuurtuin (an agricultural knowledge centre), while in 1921, four studbook bucks and four goats were imported from the Dutch province of Zeeland. Zebus were imported from British India, which were well known to the British Indians. However, the breed was difficult to milk and the animals were mainly used as draft animals (Leys 1925a, p. 412). The cost of purchasing and transporting these animals was high, but it was still budgeted for by the Department of Agriculture. Occasionally, a governor, such as Governor Staal, paid for the purchase of bulls from their own resources. Dutch peasants, who were mainly cattle peasants, also established their own 'cattle centres' on the former plantations Beekhuizen and Zorg en Hoop (Leys 1925b).

Small peasants thrived when they lived near a market. The most important market was the city of Paramaribo. McNeill and Lal (1915, pp. 170–171) commented on this: 'Amongst free Indians probably the largest class are petty landowners, while in and around Paramaribo a considerable number are cattle keepers, cart-owners, market gardeners, mechanics and day labourers. A day labourer will keep a cow or two and as he prospers buy more stock or a cart or a piece of land ... The milk supply of Paramaribo is now in Indian hands and Indians have taken over from the black population almost the whole of the land transport business.' The authors referred to the year 1911. At that time, there were 27,000 British Indians in Suriname (the total population was 85,000), 5876 of whom were under contract. As the peasantry grew, so did transport on water and land, lending (by government and private individuals), rice hulling mills and the like.

Leys (1925b, p. 450) wrote about land transport: 'In the city itself it is teeming with donkeys. Hundreds of carters from the suburbs come into town with their cart, pulled by a donkey, to try to earn something by transporting one or more cargoes for private individuals or merchants.' Photo 5.2 above shows a picture of such a cartman, indicating that all activities involving or derived from peasantry were family business. Transport by water is barely mentioned, and neither are fish catching and selling. Suriname's creeks and rivers, even the constructed and poorly maintained tributaries, were rich in fish. Some fishermen offered their wares for sale in nearby markets or tried to sell them door-to-door. As the radius of action of these fishermen/vendors was limited, sales were restricted and literally a by-catch of life in agriculture.

Photo 5.2 Two Hindustani children on a donkey cart. Transport over land became a fast-growing business, especially in and around the capital city Paramaribo

5.5 Socio-economic Differentiation

Some farms became larger than others and displayed socio-economic differences among peasants. The first form of this was found from the last decade of the last century among the 'small' Creole cocoa producers whose production was bought up by the plantations. Among them were peasants with cocoa farms of 12–20 hectares who were referred to in colonial literature as 'middle-class peasants' (De Vries 1965, p. 47ff; Heilbron 1982, Ch. VI and VII; Van Drimmelen 1918). During the first decade of the twentieth century, their share of cocoa exports was estimated at 41%. However, the prosperity of these cocoa entrepreneurs was wiped out by various plant diseases and pests affecting especially the cocoa, most notably *moniliophthora perniciosa*, described as 'petrification of the fruit' (Encyclopaedie 1917/1981, p. 190), and the slump in the economy after World War I.

A second differentiation consisted of Hindostani rice producers. During World War I, food imports stagnated. Food prices rose, which encouraged peasants to increase their production. That price rise attracted many plantation workers to the plots and caused a labour shortage (Brandon 1925). The government also urged peasants to produce food for the domestic market. The result of the increased price and the government's moral appeal was that the number of food producers rose, as reflected in the shift towards rice production and other food crops (see Table 5.3).

Table 5.3 Production of smallholder agriculture, 1900–1919, in mln. kg

Products	Averages		Percentage increase
	1910–1914	1915–1919	
Corn (maiz)	4.7	7.9	68
Bananas	1.7	2.6	53
Groundnuts	5.8	15.3	164
Rice	12.1	36.7	203
Cacoa	1.8	2.4	33

Source: Heilbron (1982, p. 279)

Table 5.4 Size of total and urban population, 1863–1923 (in brackets the percentage of urban residents)

Year	Land	City
1863	49,132	19,347 (39)
1867	49,718	20,832 (42)
1878	51,414	21,265 (41)
1884	55,533	24,758 (45)
1888	57,365	27,752 (48)
1894	64,321	29,016 (45)
1897	67,761	30,577 (45)
1903	73,542	32,585 (44)
1910	92,142	35,346 (38)
1923	112,827	44,772 (40)

Source: Colonial reports, statistical annexes

The largest growth was of rice and groundnuts, both in terms of volume and growth rate. These two products are each other's substitutes. Consumption of one should be at the expense of the other. But that did not seem to be the case: both products were among the biggest growers. That signifies that the growth of rice was not enough to (completely) replace the groundnut. Nevertheless, the leading position of rice is evident from the volume of its production and its growth rate.

The trend towards economic and social differentiation among the peasants did not continue. While the government did successfully encourage the cultivation of food crops, this was primarily for the benefit of the domestic market (K.v.K. 1916, p. 8). That market consisted mainly of the city of Paramaribo and some smaller centres and plantations in the districts. The growth of by far the most important domestic market, the city, is indicated by the increase in population (see Table 5.4).

Domestic marketing was hampered by inadequate infrastructure. Transporting of agricultural products took place over short distances by land and for somewhat longer distances by water. Ten years after the start of the First World War, a visitor would characterise the accessibility of the various places as follows: '[A] boat trip to a company on the Saramacca river took eight to nine hours, a trip to Moengo took 2.5 days. Except in the Paramaribo area, one finds only short rideable roads, from 10 to 30 kilometres in length, along the Commewijne, in and around Coronie and from Nieuw Nickerie to the Corantijn' (De Cock Buning 1926, p. 5). The

Photo 5.3 View of New Nickerie, 1905, the second-largest market in Suriname

implication of this circumstance was that peasants who lived further from the markets had to travel greater distances and had higher costs to sell any surplus. Hence, the closer the market, the higher the peasant's welfare level.

The markets were not particularly large in size, especially those located outside Paramaribo. Photo 5.3 shows the second-largest market in Suriname, Nieuw Nickerie, the capital city of the Nickerie district. The picture is of a (large) village, i.e. a limited market. Moreover, the agrarian nature of the district capitals meant that many residents engaged in farming themselves. As a result, agricultural sales had soon reached their limit.

The plots issued were mostly adjoining rectangular pieces of land. Expansion of the plot was often not possible as the adjacent plot was already occupied. Although it was possible to get a plot elsewhere, this led to the fragmentation of land where one farm could have several plots. Moreover, some areas had to be irrigated and others needed to be drained. To overcome these barriers, new agricultural areas were reclaimed and brought under cultivation in the Nickerie district in the second decade of the twentieth century. These were the Van Drimmelenpolder (1915), Boonackerpolder (1917), Corantijnpolder (1920), Henarpolder (1921) and Hamptoncourtpolder (1921) (see Annex 5.3). As an experiment, peasants paid a share in the infrastructure to be constructed, and in return they could sometimes acquire ownership rights after a number of years (Hewitt 1922, pp. 5–20; Memorie 1919, pp. 163–173; Van Traa 1946, p. 105).

Examples of the Nickerian polders recur frequently in the literature. Most of the applicants for these larger and generally more fertile tracts of land were British Indian immigrants. Their achievements were considered indicative of initiative, responsibility and cooperation. For example, on the occasion of the creation of the Van Drimmelenpolder, the agricultural teacher Hewitt wrote: 'The re-payment of contributions went beyond expectations. Many paid in full as soon as title to the land was handed over to them and many an advance was repaid early. The advance was also often not used, but the contribution was paid in cash. It would hardly have been surprising during war time, with its high rice prices, to see this unexpected solvency of those concerned (Hewitt 1922, p. 12).

The polders needed different irrigation systems. Hewitt (1922, p. 15) pointed out that '[t]he great success achieved here with the irrigation works is due to the large reservoir of water, extending from the back dams of these polders to Nanniekreek, and which may safely be estimated to be at least 2,000 H.A.' Annex 5.3 shows that in the period before the First World War, most of the polders were constructed in the districts of Suriname, Commewijne and Nickerie.

However, the quality of the soil differed. In Nickerie, the polders were often cultivated by the peasants themselves, mainly encouraged by the fertile soil and the possibility of extending existing plots (Stahel 1933). By far the majority of polders were constructed in Nickerie and Saramacca, while the districts of Suriname and Commewijne mainly featured settlements. It is also noticeable in Annex 5.3 that the population size per site, which ranged from 50 to 3668 ha, differed greatly. The small number of people per settlement site, whose houses were separated by distances equal to the length or width of the plots, and the resulting loneliness are an indication of spatial and social isolation. This can be seen in Photo 5.4, among

Photo 5.4 Bridge between the Van Drimmelenpolder and the Corantijnpolder, ca. 1920

others. Between 1925 and 1950, virtually no settlements or polders were commissioned.

Another important initiative aimed at stimulating rice production was the establishment of the Landbouw Proefstation (Agricultural Experimental Station, LPS) in 1903, which was mainly concerned with crop improvement and education through demonstrations in experimental fields (see the various issues of *West Indië* and the *West Indische Gids*; Relyveld 1954, pp. 84–96). A year later, the Cultuurtuin (Culture Garden, a knowledge centre for agriculture) was established as part of the LPS to experiment with tree species. The agricultural consultants undertook various trials, made diagnoses and visited different countries, especially British India, the Dutch East Indies, the US and British Guiana, where they oriented themselves to the crops grown in Suriname and based their solutions on comparisons with these countries.

In addition, they visited several rural areas in Suriname, where they established trials and special fields to experiment with new crops and determine yields. They had very detailed knowledge of all kinds of agricultural crops and animal husbandry as well as the beneficial use of waste products. They also had a remarkable eye for the costs and income associated with the cultivation of a crop or livestock species. Their general finding was that crops, including livestock, had to be modernised, new varieties introduced, information improved and extended, and efficiency increased. In short, the peasants' existing agricultural capital was insufficient to take production to the next level.

The transfer of knowledge was not easy because the agronomists did not always have a (good) command of the language of the British Indian peasants. Therefore they resorted to visual education: 'The main work of the agricultural department in the districts therefore consists of demonstrating, by means of visual education, the advantages achieved with selected seeds, better tools, appropriate fertilisers and careful tillage. Together with this goes the provision (against payment) of good seeds, tools and fertilisers. Other details, such as the improvement of water supply and drainage, transport to and from the fields, acquisition of land, etc., come to the fore to a lesser or greater extent' (Ref. 1921, pp. 33–34).

Credits provided by the government should also be mentioned. The need for credit for smallholder peasants was institutionalised from 1907 onwards in a *Boerenleenbank* (peasant loan banks) run on a cooperative basis (Encyclopaedie 1917/1981, p. 239). The establishment of peasant loan banks by no means meant that all categories of peasants could get loans, but it did widen the supply of money thanks to the colonial government. Occasional loans were obtained through the intercession of the district commissioner, sometimes involving a general arrangement to buy, for example, means of production or livestock (Burger 1928; Van de Houven van Oordt 1900).

5.6 Agricultural Policy 1920–1930

The period around World War I was characterised by the collapse of the export sector, which led to persistent unemployment, especially in the gold and rubber sectors. To a lesser extent, this was the case with the cocoa and coffee sectors. The malaise led to renewed interest in the plantation sector. The importance of smallholder agriculture was no longer in question, but it was considered a secondary sector. The problems there were no longer associated with the plantations as it was widely recognised that smallholders were a stand-alone sector. The smallholder sector was never an alternative to the plantation sector, however, at least not in terms of employment.

Between 1919 and 1925, at least four reports were issued that addressed the socio-economic problems of the colony (De Cock Buning 1926; Rekest 1926; Suriname Study Syndicate Report 1920; Van den Helm 1926). The recommendations were broadly the same: the diversification of exports, the improvement of infrastructure and irrigation in order to interest foreign capital, the reduction of wages or increase in labour productivity, the dissemination of agricultural knowledge, the modernisation of production techniques (mainly of sugar), the provision of tax facilities, the immigration of workers, the discipline of labour and the further exploration of the country's economic potential. The last recommendation originated from the limited success of experiments with bananas and citrus, and the booming exports of bauxite from 1922 onwards. Most of the recommendations were set aside because of the huge investments involved. That the government should bear the cost of building and maintaining the infrastructure of the settlements and polders was already seen as an indirect subsidy. But practically all reports also pointed out that peasants made a direct contribution to the government coffers.

The *Rekest aan de Minister van Koloniën* (1926) of the *Ondernemersraad* (a Council of Entrepreneurs) reflected the views of the Surinamese planters. Their main concern was the relatively high labour costs in the plantation sector, which would have been partly determined by their obligation to contribute to the costs of supply and return and the rights of immigrants. The Entrepreneurs' Council therefore called for exemption from these obligations and the establishment of a labour reservoir to control wage levels. It also stressed that infrastructure needed urgent improvement. It regretted the policy of austerity that the Dutch government had followed for the last century, since, it argued, sooner or later, due to the political turmoil in Indonesia, Dutch capital would need a reserve country for investment. Suriname should be ready for that. However, the Dutch government turned a deaf ear to the desperate Surinamese planters.

Parallel to this effort to revive the plantation sector, the expansion of smallholder agriculture led to a discussion around two related issues. The first concerned the realisation of economic potential in the agricultural sector, this time preferably with the help of Dutch colonisers (Memorie 1919, p. 168; Van Blom 1919, pp. 345–359). An earlier attempt in 1845 to colonise Dutch peasants had failed dramatically. Not surprisingly, this proposal was seen as painful and no decision was made. However, the descendants of the Dutch peasants had developed into successful cattle ranchers in the Paramaribo area. And that kept the idea of colonisation of Dutch peasants alive. The second topic concerned the use of machinery in agriculture, whose application in the United States was promising (Pyttersen 1922). The use of these machines in rice cultivation, especially the tractor, was considered useful by several experts (Brandon 1933, pp. 81–98; Liems 1921, pp. 10–14). None of these issues were discussed at length, but two decades later they were returned to.

The economic depression of the 1930s affected export markets. Coffee was virtually the only agricultural product whose exports were favourable, and the product, therefore, was supported by the government. Unemployment rose. This was exacerbated by the return of Surinamese who became unemployed in the Netherlands Antilles (Scholtens 1986, Ch. III; van Lier 1977, p. 273). Part of the unemployed population sought solace in smallholder agriculture, but there was a surplus of production even then. This was especially true for rice, which was exported in small quantities from 1929 onwards.

But rice exports were constrained by two main problems. The first was that agricultural consultants found peasants' labour productivity low and the use of resources inefficient. These issues had the full attention of the LPS. Its staff took measurements all over the country, provided information and set up trials to demonstrate the technique of new cultivation methods and rice varieties. But these found little following, partly because the results of the trials were not convincing (van Wouw 1949). The other problem had to do with the quality of the rice. The grains were too weak and broke down easily, while over the years too many varieties had become mixed together to compete effectively on the international market. Moreover, they were discoloured, which was less attractive to consumers. However, the peasant question in the early 1930s was no longer about labour supply or colonisation, but the volume of production and the quality of rice.

5.7 Conclusions

The transformation from plantation labourer to peasant took place under conditions that developed at the macro level. Those conditions consisted of the decline of plantations that set land and labour free. At least until the First World War, plantations were still important, but increasingly free labourers sought refuge on plots of land outside the plantations. This had two consequences. First, labour was expensive because the contractually agreed wage had to be paid, even in bad times. The fact that planters evaded this obligation hardly affected the obligation to pay. The second consequence was that plantations were increasingly deprived of free labour. The hoped-for competition among workers, so that wages would move downwards, did not take place because the free labourers left the plantations to establish themselves as peasants.

The historical conditions that made the second transformation possible, as well as the transformation itself, consisted of willed conditions that were sometimes the outcomes of implemented policies. Sometimes outcomes were undesired, either because policies had not been implemented or because they had failed. For instance, the government failed to save the plantation sector despite all its efforts. While the emergence of food producers for the domestic market was a spontaneous development, in which the government later took an interest, the colonial government, still focusing on the interests of plantation owners, only started to get involved with small peasants at the beginning of the twentieth century.

The plantation was seen not only as a place of work but also as an agricultural apprenticeship, but the agricultural capital of the British Indian peasants had become too outdated to take rice production to the next level. In response, the government offered new land for cultivation, provided loans, and distributed knowledge and skills related to agriculture and animal husbandry. However, the results did not meet the expectations of the agronomists, who considered the production process outdated and the resources too different. As a result, not only did rice production develop unevenly but its sale was also limited by the poor infrastructure and the sometimes long distance to the market. These are the first hard indications that peasants must have suffered from abject poverty from the early days of colonisation on. This situation must have been exacerbated by the impossibility of acquiring additional income as nearby plantations went bankrupt, and by the poor health of many peasants.

The Girmitiya peasants consolidated their position when their usefulness as food suppliers became apparent. Before World War I, they cultivated a variety of crops. The mix of agricultural products was largely known from British India or later from the vegetable gardens on the plantations. Thus, their agricultural human capital remained intact. They produced various agricultural and horticultural crops and types of meat, but it was mainly rice that came into the picture as a promising (export) product during and after the First World War. Rice had become king and sugar was hardly discussed anymore. As the colonial government hoped for surplus production, the policy effort after WWI was focused on rice, although experiments with citrus, bananas and other products continued.

During the first quarter of the twentieth century, a considerable number of official reports about the colony were issued and debated. The tone ranged from fierce criticisms of Dutch policy to pleas to pull the colony out of the doldrums. It was mainly about large government investments, improving infrastructure, large-scale immigration of workers, financing colonisation and getting foreign capital interested in Suriname. The Netherlands was denounced for its half-heartedness and frugality, as well as its lack of boldness and vision. A constant in this perspective was the plantation sector. Although the importance of the small farming sector was constantly mentioned, it remained a supplier of food and labour for the plantations. There was hardly any other focus on peasant agriculture, or any focus at all on sectors other than plantation agriculture.

References

Allen R (1999) Slaves, freedman, and indentured laborers in colonial Mauritius. Cambridge University Press, Cambridge

Ballagh J (2004) White servitude in the colony of Virginia. Study of the system of indentured labor in the American colonies. Kessinger Publishing, Whitefish (Montana)

Beck S (1924a) De Creolen en de Klein–Landbouw. In: Beck S (ed) Lezingen over Surinaamsche Problemen. Kersten, Paramaribo, pp 1–34

Beck S (1924b) Landbouwbedrijfsvorm en landbouwopleiding. In: Beck S (ed) Surinaamsche Problemen. Kersten, Paramaribo, pp 1–44

Beckford G (1972) Persistent poverty. Underdevelopment in plantation economies of the third world. Oxford University Press, New York

Bosz AJAQ, Albrecht AJ (1954) Drie Eeuwen Grondpolitiek in Suriname. Van Gorcum & Comp, Assen

Brandon HG (1925) Het Surinamevraagstuk. West-Indische Gids 6(7):481–560

Brandon HG (1933) Proefneming voor Machinale Hoeverijstculuur op de plantage Guadeloupe in Suriname in 1925–1926. West-Indische Gids 14(1):81–98

Burger JW (1928) Vergelijkend Overzicht van de Immigratie en Blijvende Vestiging van Javanen en Brits-Indiërs in Suriname. De Economist 77(1):422–445

Choenni C (2016) Hindostaanse contractarbeiders 1873–1920. Van India naar de plantages in Suriname. LM Publishers, Volendam

De Cock Buning W (1926) Rapport over Suriname, uitgebracht aan de Vereeniging van Kamers van Koophandel en Fabrieken in Nederland

de Houven V, van Oordt WJ (1900) Een en ander over kleinen landbouw in de kolonie Suriname. De Economist 49(2):599–615

De Klerk CJM (1953) De Immigratie der Hindostanen. Urbi et Orbi, Amsterdam

De Vries M (1965) De Landbouwers van Suriname. Planbureau, Paramaribo

Deerr N (1950) A history of sugar, vol 2. Chapman and Hall, London

Encyclopaedie (1917/1981) Benjamins H, Snelleman J (eds) Encyclopaedie van Nederlandsch West-Indië. Emmering (facsimile), Amsterdam

Fokken M (2018) Beyond being koelies and kantráki. Constructing Hindostani identities in Suriname in the era of indenture 1873–1921. Doctoral dissertation, University of Groningen, Groningen

Goodman D, Redclift M (1981) From peasant to proletarians: Capitalist development and agrarian transitions. Basil Blackwell, Oxford

Gowricharn RS (1985) International trade and the abolition of slavery. Caraïbisch Forum 2(2/3):70–100

Heilbron W (1982) Kleine boeren in de schaduw van de plantage. De politieke ekonomie van de na-slavernijperiode in Suriname. Doctoral dissertation, Erasmus University Rotterdam

Heilbron W, Willemsen G (1977) Transformaties in een plantage ekonomie. De politieke ekonomie van Suriname 1863–1900. Self published

Heilbron W, Willemsen G (1980a) Goud- en balata-exploitatie in Suriname: Nieuwe productiesectoren en nieuwe vormen van afhankelijkheid na 1863. Caraïbisch Forum 1(1):66–84

Heilbron W, Willemsen G (1980b) Goud- en balata-exploitatie in Suriname: Nieuwe productiesectoren en nieuwe vormen van afhankelijkheid na 1863. Caraïbisch Forum 1(1):87–101

Hewitt WE (1922) Nickerie als landbouwdistrict. West-Indië, Landbouwkundig Tijdschrift voor Suriname en Curaçao 7(4):5–20

Hira S (1982) Van Priari tot en met De Kom: de geschiedenis van het verzet in Suriname 1630–1940. Futile, Rotterdam

Hoefte R (1998) In place of slavery. A social history of British Indian and Javanese laborers in Suriname. University Press of Florida, Gainesville

K.v.K (1916) Jaarverslag van de Kamer van Koophandel en Fabrieken. Paramaribo

Kasteleyn JSC, Kluvers BJ (1919–1920). Resumé van het rapport (1919) der Studie-Commissie van het Suriname-Studie-Syndicaat. De West-Indische Gids 110:287–316

Laurence K (1994) A question of labour: Indentured immigration into Trinidad and British Guiana 1875–1917. Ian Randle Publishers, Kingston

Leys JJ (1917) Onze veeteelt. West-Indie, landbouwkundig tijdschrift voor Suriname en Curaçao 2(4):105–111

Leys JJ (1921) De Kleine Landbouwers en het Klein Landbouwbedrijf. West-Indië, Landbouwkundig Tijdschrift voor Suriname en Curaçao 6(1):31–44

Leys JJ (1925a) De veeteelt in Suriname. Nieuwe West-Indische Gids 6(1):405–418

Leys JJ (1925b) De veeteelt in Suriname. Nieuwe West-Indische Gids 6(1):441–460

Liems JA (1921) Toepassing van machinale bewerking in Suriname. West-Indië 2:10–14

Liems JA (1923) Het klein-landbouwbedrijf en de vruchtencultuur in Suriname. West-Indische Gids dl. 5(4):625–638

Liems JA (1926) Over den kleinen landbouw in Suriname. West-Indische Gids 7(1):371–384

Mandle J (1973) The plantation economy: Population and economic change in Guyana, 1838–1960. Temple University Press, Philadelphia

Mayer A (1973) Peasants in the Pacific. A study of Fijian Indian rural society. Routledge & Kegan Paul, London

McNeill J, Lal C (1915) Report to the government of India on the conditions of Indian immigrants in four British colonies and Surinam, Part II. London

Memorie (1919) Memorie van het Comité tot Bevordering van den Middenstandslandbouw in Suriname, op 29 augustus 1919 den Gouverneur Aangeboden. West-Indische Gids 1(2):163–176

Mintz S (1983) Reflections on Caribbean peasantries. New West Indian guide 57(1–2):1–17

Northrup D (1995) Indentured labor in the age of Imperialism, 1834–1922. Cambridge University Press, Cambridge

Panday N (1959) Agriculture in Surinam 1650–1950: An inquiry into the causes of its decline. Paris, Amsterdam

Pyttersen T (1922) Rapport omtrent de uitkomsten van een voorloopig onderzoek naar de levensvatbaarheid van landbouwersbedrijven in Suriname met toepassing van moderne cultuurmethodes. s–Landsdrukkerij, s-Gravenhage

Reddy M (2016) Social movement and the Indian diaspora. Routledge, London/New York

Ref. (1921) Landbouwvoorlichting aan de Britsch-Indiërs. West–Indië, Landbouwkundig Tijdschrift voor Suriname en Curaçao 6(2):33–36

Rekest (1926) Rekest van de Ondernemersraad aan den Minister van Koloniën d.d. juli 1926. Helder

Relyveld WL (1954) Mechanisatie van de landbouw in Nickerie. De Surinaamse Landbouw 2(2):75–84

Scholtens B (1986) Opkomende Arbeidersbeweging in Suriname. Masusa, Nijmegen

Shepherd V (1993) Transients to settlers: The experience of Indians in Jamaica 1845–1950. Peepal Tree Books, Leeds

Siwpersad J (1979) De Nederlandse regering en de afschaffing van Surinaamse slavernij (1833–1863). Doctoral dissertation, University of Groningen

Stahel G (1933) De Rijstcultuur in Suriname; Amsterdam, z.j. (overdruk uit De Indische Mercuur, 22 en 29 november, en 6 dec 1933), publication of the Landbouwproefstation te Paramaribo, no. 2

Van Blom D (1919) Kolonisatie met Nederlandsche Boeren in Suriname. West-Indische Gids 2(1):345–359

Van den Helm PWHG (1926) Het Suriname vraagstuk. Rapport op verzoek der Regeering uitgebracht

Van Drimmelen C (1918) Een Middenstandslandbouw voor Suriname. Uitgave van de R.K. Sociale Cursus, Paramaribo

Van Lier R (1977) Samenleving in een grensgebied: Een sociaal–historische studie van Suriname. Emmering, Amsterdam

Van Praag S (1877) Suriname: geen immigratie–leening, maar de beschaving harer negerbevolking, het krachtigste redmiddel dier kolonie: redevoeringen, gehouden in de zitting der koloniale staten van Suriname, van den 12 Februari 1877. Oliver, Amsterdam

Van Stipriaan A (1993) Surinaams Contrast. Roofbouw en overleven in een Caraïbische plantage–economie, 1750–1863. Brill, Leiden

Van Traa A (1946) Suriname 1900–1940. Van Hoeve, Deventer

Van Wouw JJ (1949) Het Departement van Landbouw-Economische Zaken in Suriname in 1945. Nieuwe West-Indische Gids 30(1):244–249

Westra P (1921) Contractimmigratie van Britisch–Indiërs. West-Indië, Landbouwkundig Tijdschrift voor Suriname en Curaçao 6(2):3–8

Willemsen G (1980) Koloniale politiek en transformatieprocessen in een plantage-economie. Suriname 1873–1940. Doctoral dissertation, Erasmus University of Rotterdam

Chapter 6
Homemaking in Suriname, 1895–1930

Abstract This chapter addresses the homemaking of Girmitiya peasants. Their homemaking equals a moral community that made it possible to mobilise the agrarian capital present in it, including the knowledge of agricultural construction, materials such as seeds, plants and animals, and the labour of group members. The homemaking suggested that cultural practices, the arrangement of nature and the practice of agriculture were intertwined. This entanglement is elaborated in Sect. 6.2, arguing that settlement as a peasant cannot be separated from experiences on the plantations because these were largely a push factor. Section 6.3 argues that survival as a peasant required not only a piece of land but also agricultural knowledge, planting materials and ties to the community. Section 6.4 examines how the Girmitiyas as peasants shaped the landscape according to their views through a specific arrangement of the geography and economy of rice farming. Section 6.5 addresses the composition and increase in smallholder production and the specialisation in rice, while Sect. 6.6 specifies this development by region.

Keywords Homemaking · Moral community · Cultural practices · Landscape · Rice cultivation

6.1 Introduction

After the end of the contract period, most of the British Indian workers settled as part-time or full-time peasants. Settlement involved making a home, known in English-language literature as 'homemaking'. This is a hard indicator that migrants settle in a new environment for an extended period of time, if not permanently. This settlement creates an attachment to the place of residence, also known as 'place attachment' (Easthope 2004). Central to homemaking is the creation of familiarity and safety, not only of the environment, but also of the house, decoration, social relations and livelihoods. Most literature on homemaking relates to settlement in urban areas, but the concept is equally applicable to agriculturists who need to make a home (Gowricharn 2020).

R. Gowricharn, *The Girmitiya Peasants in Suriname*,
https://doi.org/10.1007/978-3-031-67961-2_6

Prior to World War II, the settlement of British Indian peasants went through two phases. In the first phase (1895–1930), peasants grew various crops using their agricultural capital. As I pointed out in Chap. 4, agrarian capital was maintained on the plantations by the workers cultivating their own small plots of land, thus inhibiting or preventing any deskilling (Braverman 1974). In some cases, for example when they learned to grow new crops, that agricultural capital was even expanded. The second phase (roughly the period 1930–1950) consisted of attempts to break the stagnation of the agricultural sector. By the end of the first stage, the homemaking of the Girmitiyas had more or less been completed and enabled the operation of a moral community and the mobilisation of agrarian capital. Because of its relationship with the rice geography and economy, this chapter deals with the first phase of homemaking.

Before a discussion of the homemaking of the British Indian peasants comes the question of what the reasons were for settling in Suriname. Two motives recur in the literature on Girmitiyas (De Klerk 1953; Lal 1983, 2000; Laurence 1994). The first is that there was a thirst for land ownership, a possibility that did not exist in British India. This factor, land ownership, is given by far the most weight. The oft-quoted Dr. Comins even spoke of a passion for money and landholding for which workers would have bid farewell to their possessions in British India. The second motive for not returning was that they had crossed the *kálá páni* (literally the black water, actually the ocean). By doing so, they had jeopardised their position in the caste system and made returning to, and reintegration into, British India almost impossible. Both explanations are problematic and will be discussed in this chapter.

The Girmitiya peasants established an ethnic group through which they felt culturally at home. Ethnic group formation as cultural homemaking made it possible to mobilise the agrarian capital present in it, including the knowledge of agricultural construction, materials such as seeds, plants and animals, and the labour of group members. This, in turn, made it possible to recreate the physical environment and to build an economic livelihood by growing familiar crops. Cultural practices, the arrangement of nature and the practice of agriculture were thus intertwined.

This chapter addresses the homemaking of Girmitiya peasants. That process largely involved an analytically distinct cultural, geographical and economic home, all of which were closely intertwined and enabled the establishment of farms with rice as the main crop. Peasants lived in small concentrations and relied heavily on the agrarian capital, which was in fact ethnically specific capital, for rice cultivation. The development of rice cultivation meant a reshaping of the landscape, with rice fields and cattle farming as the most recognisable elements. Although the government tried to make rice an export product, that policy was not very successful until World War II. The production capacity of the average farm was limited by the paucity of available labour. Put differently, the economic home was severely constrained by the cultural home, especially by the low degree of family formation.

In the next section, I will elaborate on the workers' motives for settling in Suriname, arguing that settlement as a peasant cannot be separated from experiences on the plantations because these were largely a push factor. This causality has hardly been addressed in the historiography of Girmitiyas. Survival as a peasant required not only a piece of land but also agricultural knowledge, planting materials and ties to the community that together determined agricultural capability. That capability is discussed in Sect. 6.3. Section 6.4 examines how the Girmitiyas as peasants, given their capital, shaped the landscape according to their own views through a specific arrangement of the geography and economy of rice farming. Section 6.5 addresses the composition and increase in smallholder production and the specialisation in rice. Section 6.6 specifies this development by region. The concluding Sect. 6.7 interprets the findings as homemaking of Girmitiya peasants.

6.2 'Suriname, This Country I Have Chosen'?

In 1968, Hindostani poet Srinivasi (Lutchman) wrote a poem titled 'Suriname, this land I have chosen'. It is a nationalist statement expressing the choice between two countries, with in this case a preference for Suriname. The choice of words suggests that it is a decision made freely and consciously and not (partly) motivated by restrictive circumstances, temptations or fears. According to a widespread explanation, British Indian workers in almost all plantation colonies would have been motivated to settle in the new country by the prospect of having a piece of land at their disposal. Land ownership was, for the vast majority of workers, a dream that could be realised in the country of settlement. Thus, Dr. Comins (1892, pp. 11–12) wrote about the passion for landholding of the British Indians. Similar explanations for settlement by other authors can also be found (Bisnauth 2000; Choenni 2016; Emmer 1985a, b; Ramdin 2000). While this explanation cannot be entirely discarded, a few remarks are in order.

In 1878, after the first batch of workers had served their contracts, a small number of British Indian immigrants settled on domain land. Formally, the immigrants waived the right of free return, but in practice this option was kept open. If they waived the right of free return, they received a premium of f100. The workers could settle in government-established settlements, which De Klerk (1953, p. 163) described as '... abandoned plantations, returned to the bosom of the domain, or bought by the government and parcelled out into plots for the benefit of the smallholders ...' The government took care of deforestation, roads, drainage and the like, but these plots could not be owned; instead, the peasants had to pay rent. But the enthusiasm for settling as peasants was not great then.

In 1895, the possibility of owning land was opened up, after having farmed it satisfactorily for a number of years. The British Indians lost the right of free return only if they accepted the premium of 100 guilders. So they remained in Suriname because they could still return to British India. It was this possibility, the preservation of the right of free return time, that was the main motive for settling in Suriname. This was evident from the rapid increase in the number of land users after 1895. In 1892, 1900 and 1904, the number of land users was 172, 828 and 3542, respectively (Burger 1928, p. 428). Based on this increase, Burger correctly argued that the colonisation of British Indian workers dates from this period. After that, various settlements were established, such as Paradise (1895), La Rencontre (1897), Hecht en Sterk (1899), Laarwijk (1899) and Nieuw Waldeck (1902). Annex 5.3 lists the settlement sites, arranged by year of establishment and district. Most settlement sites were set up in the districts of Suriname, Nickerie and Commewijne.

The decision to settle as a peasant may also have been motivated by the desire to leave the plantations as soon as possible. In political history, this behaviour is known as 'voting with the feet', a term for making displeasure or disapproval clear by sometimes literally walking away. The origin of the term cannot be precisely determined, but most authors believe it was first used in 1917 by Vladimir Lenin in his analysis of the desertion of Russian soldiers. Since Lenin, the term has been used in various contexts to interpret behaviour. In the Caribbean, the departure of former slaves from plantations has been interpreted in these terms (Mintz 1983). For Suriname, van Lier (1977, p. 182ff) gave a similar explanation for the behaviour of Creole workers who left agriculture to work in gold and balata mining.

A second explanation for settlement in Suriname is that of the *kálá páni*. According to this narrative, those who crossed the ocean lost their position in the caste system. De Klerk (1953, pp. 88–89) explains that the term originally referred to captivity, but Nayar (2015, p. 97) believes the concept was created in the colonial period. 'It refers to the metaphoric black waters of the Atlantic which indentured Indians … crossed en route to the Caribbean. The term therefore is irreducibly linked to colonialism and indentured labour.' The link to the Caribbean is inaccurate because authors on Fiji and Mauritius also refer to the *kálá páni* to avoid going back to British India (Kumar 2017, p. 244; Lal 2000, p. 122).

The *kálá páni* statement was originally launched by Comins (1892, p. 17), who wrote about the labourers: 'Their belief is that no man calls himself a Hindu who has crossed the sea, so they lose their respect for the caste and religion of their fathers which they neglect, and acquire no other in their place. They still wear amulets and charms and believe in the evil eye, ghosts and devils innumerable, but in no God. One of the reasons for not returning to India is that they would be despised and mobbed in their villages, or have to spend much money for their re-admission to their caste.' Some returnees have been disowned for this reason, while others have not ventured to return for fear of failure (Kumar 2017). As an explanation for not returning and settling in the new country, the *kálá páni* reasoning is unconvincing (Gowricharn 2023). For about a third of the workers did go back and by so doing contradicted the alleged ban by the *kálá páni*. Moreover, those who settled in Suriname did not do so for fear of the consequences of returning, but for various other motives.

Formally, the workers had the option of returning to British India after the contract expired, but they preferred to settle as peasants without loss of rights. This preference suggests that the former indentured labourers wanted to supplement their savings—if they had any—with agricultural activities. It was in all likelihood after the end of the contract that the idea arose among the peasants to settle permanently in Suriname (see Fokken 2018, pp. 203–204). Based on quantitative data, Kraijo (2022) came to a similar conclusion. For the period after indenture, familial relations, especially those formed in Suriname, appeared to have a strong effect on the relative share of Hindostanis settling. The peasants were not faced with a choice between British India and Suriname as Srinivasi's poem suggests. The idea that this was a choice suggests that there was a conscious and rational consideration. There are no indications of that.

A major issue here, however, is whether the experience on the plantations was so negative that it was a reason for workers to seek refuge elsewhere. This need not apply to every worker. As Chap. 4 showed, there were different categories of workers, with different abilities and different opportunities. But there was definitely a tendency to 'flee the plantation'. The figures of Emmer (1984, p. 233) and Fokken (2018, Appendix 3.5) show that the flight of labourers declined around the transition from the nineteenth to the twentieth century. This behaviour was probably related to the possibility of fleeing to the settlement places and other lands that were opened up at that time. This explanation presupposes that life on the plantations was intolerable, a push factor, hence flight. Settlement as a peasant can therefore be understood not only from the perspective of the desire to own land but also from that of the wish to escape plantation labour.

There are additional reasons to take the interpretation of voting with their feet seriously. Hoefte (1998, p. 127) points out that, on average, only 194 British Indian workers renewed their contract each year, while among Javanese the number was almost 1000. She concluded that '...Hindustanis preferred to settle away from the plantations'. Emmer also came to a similar conclusion. He found that 34% of men entered into a re-contract and 14% of women. 'A second labour contract was usually considered by most Indian immigrants as a last remedy. As soon as it was even possible, Indian labour migrants, both men and women, returned to India with their savings after the end of the first contract period or opted for permanent residence in Suriname, but as small peasants or free labourers' (Emmer 1984, p. 222).

The voting with their feet interpretation is further supported by observations by Dr. Comins. 'Since 1873, 325 indentured or contracted immigrants have commuted the remainder period of their indenture by payment of a portion of the indenture fee equal to the part of the indenture expired, and in the case of "contracted" coolies part of the premium paid them, and this way have freed themselves from their engagement' (Comins 1892, p. 31). This may have taken place for various reasons, including poverty and humiliating treatment on the plantations.

However, two authors make positive judgements about the earnings on the plantations. Choenni (2016, p. 525) believes that the wages earned were so high that workers were able to save about 25 ct (i.e. about a third) of their daily wages, and Emmer (1984, p. 215), writing about women workers, argued: 'In many cases, the income of their male partner was so high that the housewives did not have to take part in the heavy plantation work.'

Comins (1892, p. 32) points to another reason that compelled workers to settle in Suriname. 'The position of those who, having declined further contracts, leave the estate and wait for the ship is not an easy one, and if this period is prolonged until their means are exhausted and there is still no news about the ship, there is little for them to do but to return to the estate or to be sent to the Poor Asylum.' Note that Comins speaks in the plural, so it was not a single case. You will recall that Leys

(1925) had pointed out that during World War I, many workers were unable to return to British India because of the fact that shipping was virtually at a standstill, and workers settled as peasants for that reason. That may have been a provisional decision that became permanent over time.

The spreading of negative reports about British India would also have encouraged workers to stay in Suriname. Similar instances have also been reported by Bhagwanbali (2010) and Hassankhan (1993). The latter reports: ‘Not everyone made a voluntary choice in doing so (staying, Author). First, it may be noted that some of the indentured labourers in Suriname died before the expiry of the contract. In addition, there were not always opportunities to return at the time the people involved wanted. Return was therefore postponed until one could not return because one did not have the money to do so’ (Hassankhan 1993, p. 17).

Fokken (2018, pp. 203–204) additionally points to other reasons that prevented return. The first was that passengers were required to pay for clothes and other necessities. Moreover, it was not possible to return every year. This was the case, for example, in the years 1906, 1908 and 1910. The long waiting time not only depleted savings but also discouraged workers from returning. In addition, the head interpreter Sitalpersad, who held a high position in the Surinamese administration and often used his influence at will, personally made it impossible for the indentured British Indians to await their departure at the coolie depot. And finally, various parties, including Sitalpersad, the Suriname Immigrants Association and colonial authorities, spread the rumour that the situation in British India had become unfavourable. Fokken (2018, p. 204) therefore concludes that the return was discouraged by fellow group members and authorities.

Given these circumstances, the question is to what extent settlement in Suriname was a free choice or determined by the passion for land or by the crossing of the *kálá páni*. Freely made decisions cannot be ruled out, but one may wonder why workers who left to earn money and planned to go back did not do so. The passion for money and landholding argument, if this motive was widely held, was by no means the only motive for settling in Suriname. It should also be pointed out that the decision must have ‘matured’ over a longer period of time. After all, immigrants could settle as peasants for a long time without losing their right to free return. Various individual circumstances may also have contributed to settling in Suriname: the desire for partners, the presence of children, new friends and relatives (Kraijo 2022). In short, the decision to settle in the colony was probably determined by a complex of factors and not because things were so much better in Suriname than in British India.

6.3 Homemaking

Homemaking is largely based on nostalgia and thus an aspiration to recreate the situation as one knew it from ‘home’ (Blunt 2003). Nostalgia is undoubtedly a driving force in the process of homemaking, certainly for the first generation, although various local forces also influence the process. Homemaking by peasants, for example, is influenced by weather, vegetation, animal life, seasons and soil conditions (summarised as the natural environment or geography). Peasants also develop the products they know from home, whether in a modified form or not, to build a living in Suriname. Knowledge of all these aspects is local-specific and is known as 'local knowledge' (Geertz 1983/2000).

The British Indian peasants knew, analytically, three forms of homemaking: cultural (the formation of an ethnic group), geographical (the rearrangement of the geographical environment) and economic (growing familiar products). The cultural and geographical forms of homemaking were decisive preconditions for economic homemaking. The cultural form enabled peasants as an ethnic group to put customs, knowledge, materials and relationships to use in farming with the help of relatives, neighbours and friends. The geographical form, which consisted mainly of the cultivation of plants and animals and the (re)arrangement of the physical environment, was also determined by 'home knowledge' and the availability of 'ethnic labour'. This disposition of knowledge, labour and material depended on a moral community that could be mobilised when some stages of agrarian cultivation required a collective effort (Carrier 2018). More specifically, thanks to the moral community, cultural and natural resources could be developed and a rice economy emerged.

6.3.1 *Cultural Homemaking*

As previously mentioned, cultural homemaking amounted to the formation of an ethnic group. By developing relationships, as had already been done at the depots in Calcutta and Paramaribo, during the journey to Paramaribo and on the plantations, British Indian peasants developed an ethnic group and made a cultural home. At the scattered settlement sites, small ethnic communities emerged based on slumbering ethnic ties that were characteristic of a quasi-group (Mayer 1966). This kinship was expressed, among other things, by a racial resemblance, by common descent (British India as a frame of reference) and kinship in terms of language, religion and morals and customs. Group formation took place through the amalgamation of the diverse population, although this unification was not total because of, for example, differences in religious affiliation (Hindus and Muslims) or caste differences that still persisted (Kruijer 1949). The beginnings of amalgamation could be spotted quite early, such as in *dipuá* and *jaháji* relationships and mixed marriages. Whatever small groups had formed on the plantations, they grew into the larger Hindostani community on and outside the settlement sites (Gowricharn 2013).

The ethnic group consisted of relationships between families, and more broadly between peer groups, supported by communication, working with community norms, recognition of moral incentives and obligations and interaction in everyday life. Without the common understanding of moral rights, obligations and incentives, it is impossible to supplement the labour of an individual household with additional labour, for example from neighbouring households or from family and friends (Carrier 2018). Ethnic relations are not an obvious concept when discussing economic cooperation, since economic behaviour is usually seen as individual and governed by material incentives (Stiglitz and Walsh 2006). But such cooperation is well known in the case of important events such as funerals, marriages and the organisation of community festivals. Due to this moral community, the borrowing and giving of agricultural capital—seeds, cattle, plants and the like—and the labour of ethnic group peers could be enacted.

Ethnic cohesion was noted in colonial literature, but not connected to the labour process, at least not explicitly. In the first quarter of the twentieth century, agriculturalists and other reporters often referred to specific characteristics of ethnic groups: British Indians, ex-slave population, Javanese and Dutch peasants (see, for example, the writings of Beck, Brandon, Burger, Leys and Liems). In doing so, they implied ethnic-specific group performance. Brandon (1925, pp. 487–488) wrote about the British Indian peasants: 'Indeed, these people had cleared their land neatly, drained it properly and planted it well. But these people were peasants, had an apprenticeship in drainage of polder lands of at least five years behind them on one plan or another, and had been helped in their reclamation work, according to statements by the supervisor, by friendly compatriots.'

The formation of local ethnic groups is important for mutual assistance during the agricultural process. The size of the allotted plots was based on the estimate that a family could make ends meet with this for 3 days a week and had to go to work on neighbouring plantations for the remaining income (Burger 1928). But this policy failed, not only because many plantations closed but also because there were few families. As outlined in Chap. 3, the ratio of men to women, a purely numerical indicator, remained unfavourable until after World War II. Burger (1928, p. 440) calculated that in the period 1905–1910, there were 444 women to every 1000 British Indian men, and in the period 1910–1915, there was a smaller number of women, 355 to every 1000 men. What caused this decline is not clear, but there could be no question of family formation on a large scale. It is therefore strange that in Hindostani circles the narrative has taken root that the joint family was the characteristic family form (Adhin 1960). If the joint family developed in Suriname in later periods, it was only possible when the demographic balance had recovered somewhat and plots were large enough to support several families.

Community formation was determined not only by quasi-group and latent ethnicity forces but also by the concentration of British Indians in the various localities. In these residential concentrations, new forms of assumed kinship developed, known among Hindostanis as *gáwn ké palwár* (kin of the village), a point that Choenni (2016, p. 618) drew attention to. The concentrations allowed for further expansion of rudimentary ethnic institutions. Various forms of kinship and internal mixing could crystallise, such as in language, food, moral obligations, manners, religious customs, feasts and other rituals (Choenni 2016; Fokken 2018; Gowricharn 2013).

The trend towards merging and expanding communities was also based on racial and cultural similarities. In 1903, Agent-General Van Drimmelen reported that British Indians were aloof from Creoles, even though they possessed the knowledge of that group and had a command of their language. De Klerk explained this attitude from a self-image perspective: according to him, British Indians rejected assimilation because they preferred to manifest themselves independently as a group, and they also feared that 'racial uniformity' would not turn out to their advantage, because, among other things, it would hinder their ability to perform (De Klerk 1953, pp. 209–210; see also De Klerk 1941). The internal involvement was high, and so was the aloofness towards other groups.

Local group formation made it possible for agrarian capital to be activated and applied for the benefit of rice cultivation. In other cases, when an individual had left for Suriname as a child from British India or had been born in Suriname, agricultural knowledge had to be transferred. It was also possible to enlist the labour of neighbours and relatives who possessed the same knowledge. This applied not only to assistance in the production of agricultural crops but also to the care of domestic animals (chickens, ducks, goats, sheep and cows). However, enlisting the help of ethnic peers was sometimes not optimal because roads did not allow it, there were not enough people, the health of neighbours did not permit such help, and there were sometimes quarrels and envy among neighbours (Kruijer 1949, p. 618).

In addition to the conducive and restraining conditions for cooperation, spatial distance was a highly neglected condition. Photo 6.1 provides a glimpse of this.

The area of Kwattaweg was already known as a horticultural area around 1910 (Beck 1924b; van Lier 1977, pp. 174–175) and thus was relatively densely populated. Yet the photograph shows that the peasants lived far from each other, so residential isolation must have been high. At settlement sites, peasants sometimes lived

Photo 6.1 Kwattaweg around 1920, a horticultural area near Paramaribo

in clusters that varied greatly in size, which can be seen from the size of the number of hectares in Annex 5.3. The photo shows that houses were located at relatively large distances from each other. They were minimally separated by the size of the plots (1.5–2 ha). This was enforced to a certain extent by the ribbon development (Van Dusseldorp 1963). Before the Second World War, only the leaseholders in the settlements lived closer together. Outside the settlements and in the polders, the distances were greater because the plots were larger.

After the war, the space between the houses was 'filled in' by the construction of new houses, thus reducing the residential isolation. Moreover, the roads were improved and water pipes and electricity were gradually installed. As a result, the polders and settlements took on the character of an urban neighbourhood. This is especially evident in the Nickerian polders, which incidentally did not prevent the steady urbanisation from that district. In the Saramacca district, however, many settlement sites, polders and villages were abandoned. Only the names of places remind us of bygone communities. Also, in the districts of Commewijne and Suriname, many residential concentrations have been abandoned or, by contrast, urbanised over the years.

6.3.2 *Geographical Homemaking*

The second form of homemaking involved the rearrangement of the surrounding geography. The settlement of British Indian peasants meant that they had to arrange the natural environment. This was done on the basis of knowledge, customs and traditions they knew from British India. The main activity that shaped the landscape was the deforestation of overgrown plantations, some of which had been abandoned decades ago. The hard labour required for deforestation has been described in detail by Beck (1924a, b), Leys (1921) and Liems (1918), among others. Deforestation was mainly a physical effort that is difficult to overestimate.

A second major change in geography involved the creation of sawas (paddy fields). Here too, both the agrarian capital and the moral community were indispensable. After a visit to former British Guiana in 1891, the British Protector of Immigrants, surgeon D.W. Comins, reported that the British Indians cultivated rice 'in exactly the same way as in India' (cited in Bisnauth 2000, p. 154). Referring to Indian villages, Bisnauth (2000, p. 25) described the cultivated land as having 'various boundary markings; these were earthen ledges that also served to retain rain or irrigation water'. British Indian peasants' designs for houses, stables and paddy storage were also the same (Bisnauth 2000, pp. 152–153). The immigrants destined for Suriname and British Guiana were recruited from the same agricultural communities (De Klerk 1953; Laurence 1994). Their common agricultural capital was reflected in the similar way in which they organised their environment in British Guiana and Suriname.

Especially on plots of land outside the settlements, the land had to be rearranged and cultivated. In the first place, this involved deforestation, a job that required a lot of physical labour, because in those days tilling was only done with wooden tools, axes and wooden spades. Tree stumps and roots had to be removed and the soil then excavated. The soil was then ploughed and harrowed before rice could be planted. The heavy clay soil was fertile, fed naturally by rotting vegetation and immersed in rainwater. As in British India, rice fields were laid out and divided into rectangular cultivation plots, demarcated by constructed dams. The drainage systems were supported by canals and gutters (Stahel 1933). An important task for the beginning peasant was the drainage of water in the rainy season and irrigation in the dry season. The ploughing of soil softened by water was greatly facilitated by the use of draft animals, usually cows and oxen (Liems 1926; Stahel 1933). This construction was also very labour-intensive.

A third element of geographical homemaking was the keeping of cows, goats, chickens, ducks and sheep. Most cows were kept for milk and for their usefulness in ploughing the land, threshing and transporting paddy. Cattle farming required stalls, good feed and care. The use of animals in rice production was therefore something only families with suitable land and financial resources could afford. The colonial administration took the initiative in importing, breeding and selling bulls as draft animals. Peasants could buy these animals with their savings or with loans obtained from the government (Burger 1928; Leys 1925). The use of cattle, however modest, underscored that the vast rice fields could not be cultivated by manual labour alone.

The size of the plots, which can be clearly seen in Photo 6.2, also meant that houses were situated at relatively long distances from each other. At a time when distances were great for people on foot and electricity and water supply were not

Photo 6.2 Rice fields in the district of Commewijne, 1947: part of the geographical homemaking

available, farming must have been a lonely existence for people in poor health. The picture also shows the landscape as developed by Hindostani peasants. This used to characterise large parts of agrarian Suriname but has been destroyed by project developers over the decades. It is a peasant-cultivated landscape that is part of Suriname's heritage.

The landscape was also characterised by the specific construction of houses, which was not only the result of a concerted effort but also contained ethnic profiles. Photographs of houses taken in the last decades of the nineteenth century and the first decades of the twentieth century show that these houses were small huts whose floors and walls consisted of a mixture of wood and were cemented with a mixture of mud and cow dung. The housing was almost symbiotic with the surrounding vegetation.

Some houses of the British Indian peasants could be recognised as Hindu dwellings by the many *jhandies*—bamboo poles with flags of different colours that were part of Hindu rituals. The familiarity of the yard was underlined by the trees, animals (poultry, goats, sheep) and vegetation around the peasants' houses (Beck 1924b). Until the 1970s, the peasants' farmyards were still recognisable. Trees were planted around the houses. Some fruits from these trees, like jack fruit and bananas, served as food. In addition, the yards of Hindostani peasants were planted with various vegetables (some of which require climbing frames, such as long beans), tomatoes, peppers, and larger plants and trees, such as the neem, katahar, sapotille, banana and jack fruit. The vegetables are found in close proximity to the house, while the larger plants and trees are somewhat further away (Gowricharn 2016, pp. 120–121).

Haraksingh (1999, p. 40) refers to similar changes in the landscape in Trinidad in the early stages of British Indian settlements, specifically to the trees and other vegetation around the houses, temples and mosques, the breeding of animals, the construction of ponds and tanks, the diversion of water sources and the clearing of land. He concludes that 'Indians had begun to think of Trinidad as their home long before general opinion in that country had awakened to that possibility'.

6.3.3 Economic Homemaking

Economic homemaking was enabled by cultural and geographical homemaking. It included the cultivation of crops for sale and for personal use. The latter should be understood more broadly than own consumption. Many proceeds of plants such as fruits and vegetables were eaten, while others served as decoration (flowers). Other plants were used as medicine, in religious rituals or during festivities. Practically all ethnic groups had their own set of plants for consumption in this broad sense. This field of study, known as 'ethnobotanica' (Van Andel and Ruysschaert 2014), is emerging. With the exception of Trinidad, this topic has hardly been addressed in the Indian diaspora. An exhaustive survey of 'Hindostani flora and fauna' is lacking and usually claims are little more than plausible (Mahabir 2004, 2008; Samaroo

2017). The field is complicated by the fact that the origin of many plants and animals is unclear as they allegedly originate in both Africa and Asia, and moreover because Hindostanis learnt the use of some (especially medicinal) plants in Suriname from Creoles and Javanese (Raghoenandan 1994).

A study often referred to describing traditional peasant life is Bihar Peasant Life (Grierson 1885). This study is highly lexicographical and contains descriptions of all kinds of aspects of peasant life in Nagri script (the script in which Hindi is written) and in English. The book details the labour process and shows how much it is surrounded by rituals. The economic activities of British Indian peasants in Suriname were not framed in rituals. Grierson's overview does disclose, however, that tillage, rice planting, sowing, harvesting, threshing and hulling, as well as the use of animals in the labour process, show great similarities to what is known in Suriname (Grierson 1885, pp. iii–v). This gives support to the view that the geographical home in Uttar Pradesh and Bihar was largely reflected in Suriname.

A list of the best-known medicinal plants and vegetables provided by the Trinidadian historian Samaroo is included in Annex 6.2, supplemented by findings by the Trinidadian anthropologist Mahabir (2004, 2008). Based on their overview, the vegetation can be divided into six categories, as listed below:

- Agricultural crops: these include especially rice that the British Indians brought to Suriname or the Caribbean (Stahel 1933).
- Vegetables: the Girmitiyas also practised horticulture, especially the cultivation of vegetables. An ethnobotanical study confirmed that numerous vegetables grown by British Indian peasants originated from the former British India or from South-East Asia (Raghoenandan 1994).
- Fruit: well-known examples include the mango, djamun, jackfruit and bananas that now characterise the agrarian landscape. Fruit trees often grew in peasants' yards, i.e. in the circle around the house.
- Medicinal plants: neem [read niem] and dhanyá are medicinal plants. They sometimes grow in the wild and sometimes in farmyards. The neem is a tree whose leaves are used medicinally.
- Flowers: gená, also known as africana, is used in social and religious rituals and to make wreaths (see Goerdayal 2019 for some other examples).
- Herbs and plants: podiná, a plant whose leaves are used in Indian cooking, especially for making chutney.

The use of many of these and other plants is culturally specific. Examples include banana leaves, which were used to serve meals during Hindu weddings and other religious ceremonies at the time. The leaves of mango trees are still used in Hindu religious rituals, as are gená (africana) and tulsi (basil). These examples of culturally embedded botanical use among British Indians can easily be extended (Goerdayal 2019; Mahabir 2008; Raghoenandan 1994; Samaroo 2017).

How did these ethnobotanicals end up in Suriname? Samaroo (2017) argues that the immigrants brought all kinds of plant material in a *jaháji* bundle, which he translated as 'ship belongings'. One of the major problems in this field of research is that while sometimes the origin of the plants and animals can be shown (Van Andel and

Ruysschaert 2014), it remains unclear whether the immigrants themselves brought this material to Suriname. Nowhere is it recorded whether they took plants, seeds or animals with them. It can also sometimes be made plausible that a certain crop originated from South-East Asia, but that does not exclude the possibility that the crop already existed in Suriname before. Sometimes this involved a different variety of the crop.

A second mode of dissemination is said to have taken place through the British Royal Botanic Gardens. This research institute set up botanical gardens in the colonies, such as the Botanic Gardens in St Vincent and Trinidad. Part of the research involved transporting plants from one place to another. Conversely, departments in the Caribbean imported plants and animals from other parts of the British Empire. This explains the spread of plants and animals from Asia to the Caribbean even before British Indians had set foot there. Occasional dispersal occurred due to the British bringing hijacked cargoes to other British colonies. This would have been the case, for example, when the British seized a French ship carrying mongooses. These animals were used to kill rats and rattlesnakes on sugar plantations, but they later became a pest because they feasted on chickens and ducks (Samaroo 2017).

The assumption of transport of seeds and plant material in a *jaháji* bundle has important implications. If the agricultural material had been brought by the immigrants, they prepared themselves for a prolonged absence and even settlement in another country. In that case, it is unlikely that they were (widely) misled in their recruitment as labourers. However, it is not ruled out that requests for agricultural materials were sent through the return migrants, but there are no indications of this.

The possibility also exists that the government imported the agricultural material as McNeill and Lal (1915, p. 172) suggest: 'Government not only helped the sufferers by loans but supplied seeds and plants from Government nurseries to those who wish to replant their plots ….' Analogous to the distribution of plant and animals by the British, the Dutch too obtained numerous sugarcane and rice varieties and fruits from the Dutch East Indies as casually mentioned by government officials. But not only from there. Agriculturalists in Suriname also regularly kept abreast of developments in the western hemisphere, especially the United States and British Guiana. In 1947, they established the National Herbarium, a botanical institute that increases and manages plant collections. Its function, especially compared to the Royal Botanic Gardens, is limited. The point to emphasise is that the plants and animals were in all likelihood distributed mainly through colonisers' institutions rather than immigrants.

It is also plausible that some plants and animals, or variations thereof, were already present in the Americas before the English and Dutch had spread them. Botanical elements from abroad are sometimes easily replaced by local plants, especially vegetables, fruits and flowers. The relevance of the origins of plants grown by British Indian peasants should therefore not be overstated. The origin of the plants and animals was only relevant when the peasants were familiar with their use. If this was not the case, it is more likely that the peasants learnt about the cultivation and use of plants and animals from other ethnic groups. The importance of ethnobotanicals, i.e. plants that are historically or in terms of use specific to a particular ethnic group, is that they have been incorporated into the lived culture (as food and other uses), as commercial crops and as a specific feature of their physical environment. By cultivating the crops on a large scale, peasants shaped the physical environment, changed the composition and extent of the vegetation, and thus created a familiar landscape as part of their habitat.

Little is known about the use of animals. It can be inferred from casual observation by agronomists that most animals were imported at a later date by the government, at the request of peasants, planters and agronomists themselves from British India, the Netherlands and Java. This was especially true of zebus (bullocks), which were used as draft animals and were well known in British India. Mahabir (2004) argues that in addition to zebus, the British government also transported several varieties of buffalo (both as draft animals), mongooses (to control rat infestation), goats and monkeys (sometimes via London and other intermediate stations) to the Caribbean region. Plants, trees and animals formed a decisive part of peasants' economic capital in the agrarian environment, alongside land, knowledge and tools. This enabled them to create a familiar physical environment, make a living and sell products.

The three forms of homemaking, the cultural (the formation of the ethnic group), the geographical (the rearrangement of natural resources) and the economic (growing crops for own use and for sale), were closely related. All three forms had their origins in the knowledge, traditions, customs and labour embedded in the ethnic group. As a result, rice cultivation led to a cultural geography and a cultural economy that belongs to the heritage of Surinamese Hindustanis. This landscape is still observable in the 2020s.

6.4 Agricultural Productivity

Because culture and economy were closely intertwined, homemaking could not take place in isolation from the forces of production. It is therefore important to look more closely at the distribution of factors of production and the productivity of these factors. Land or labour, for example, may have been available, but how productive they were differed by area and household. The economic capability had a direct effect on the cultural and geographical home, and, conversely, the cultural and geographical home affected the economic home. If this connection was not present, peasants had great difficulty in mobilising labour from kith and kin and neighbours, setting up the environs and building an economic livelihood.

The British Indian peasants, when they started out, had various degrees of access to factors of production such as land, labour, capital, raw materials, technology, knowledge and management. Remember that some workers had been able to save, which gave them start-up capital, while other peasants hardly possessed capital of any significance. Knowledge of agriculture could also differ. In particular, peasants who had come to Suriname as children or who had been born in the colony hardly possessed any knowledge of agriculture. What they knew they had learnt on the plantations or at a settlement such as Alkmaar. The obtained land was also of very uneven productivity, as Annex 7.5 shows. The amount of labour varied with family size, but families were few and far between, especially in the early period. Moreover, many people were sick and weak, which greatly affected productive capacity. And finally, it should be pointed out that the location of the land mattered. If the land was located near an outlet, there were opportunities to sell the surplus produce. Otherwise, there were not. Savings, knowledge, land, labour, health and location determined the productivity of the peasant household.

The wide variation in production factors rendered production unequal per farm and per region. Leys (1916a) argued that land productivity faced two major problems: tillage and fertilisation. The first problem arose from the shallow tilling of the soil, patrolling and the levelling of the soil for planting. This made the clods even smaller, which made the soil compact and—the longer it was used for rice growing—gave it the character of heavy clay soil. The second problem was the inadequate fertilisation of the soil. The weeded grass, rotting plants and decomposed wood, summed up as humus, contributed to fertilisation, but in order to do so, this material had to be strewn over the rice fields. Many peasants burnt this material, wasting precious manure. 'The weed that grows on the rice fields, the straw left over from the previous rice fields or the wheat, which we grew as a midday product, cost us nothing. Only the mowing and ploughing cost us some effort' (Leys 1916b, p. 1).

In several publications, Leys discussed the poverty of the small peasant, focusing mostly on the Creole peasant. The peasant's suffering was down to themself, he believed, because they were not efficient, either in terms of land use or crops grown. Leys' view of the British Indian farms was a bit sunnier: they would manage, provided the government created the conditions for serious business. But own fault or not, there was abject poverty among all categories of small peasants. That is why Leys advocated the development of middle-class agriculture. I let him speak at length because his advocated solution differed greatly from the government's colonisation concept that consisted of allocating 1.5–2 ha of land in the vicinity of plantations.

'Now is the time to seriously pursue the development of the smallholding into a medium-sized farm. A farm of 5 HA can be operated by a family with little help from outside, if one sets it up properly. Two HA of pasture. One HA for growing fodder as supplementary feed for cattle. Paragrass, green maize, green millet and some leafy leguminous plants can be used for this. One and a half hectares for the cultivation of market crops: cassava, potatoes, napi, maize, ochre, rice if the condition of the land allows it, bananas and plantains, and capuchin and beans. Half a HA for the house, the stable, and the vegetable garden and fruit trees. The order of the parts can be changed as required in the following year. On such a small farm, two cows can be kept, which can give at least 2000 litres of milk per year, and produce on average one calf per year, which can be sold.

'Furthermore, an average of five pigs can be supplied as mature fattened animals for slaughter, while at least two more can be slaughtered for own consumption. If you want, you can also breed four or five goats for meat, milk or trade. One or two donkeys will have enough to eat. Chickens and ducks, if there is water, provide a nice extra income' (Leys 1921, p. 43). That, according to Leys, was the solution to the poverty problem. Practice differed greatly from the idea that farms had to be 5 ha in size. Table 6.1 is telling.

First, a discrepancy in the figures. The size of the farming population in Table 6.1 differs from that in Table 5.2. In Table 5.2, the size of the agricultural population in 1910 was about 39,621, while according to the above table it was 29,328. The difference may be due to the inaccuracy of the data, or more unkindly phrased, to the unreliability of colonial statistics, which van Wouw (1949, p. 248) formulated as'a 'hectic gradient of figures'. But it could also be that Table 5.2 shows the total population (i.e. employed and children) and Table 6.1 only holders of plots. In the latter case, that would mean that every holder had at least one plot. But that could not be the case as the number of plots is smaller than the population. The difference remains a mystery, which incidentally is more common in colonial statistics (Gowricharn 1991, Ch. 2.6).

In three cells of Table 6.1 (see the table at the bottom), there are asterisks by numbers that differ significantly from the other figures and, in my view, indicate misrepresentations. To minimise distortions, I took the average of the 'correctly stated 'numbers and used that as an approximation of the correct number. The

Table 6.1 Population, numbers of plots, hectares cultivated in small farming on settlement sites and outside, 1906–1930

Year	Number immigrants	Number non-immigrants	Total population (1)	Number parcels (2)	Total area (3) Between brackets (3/1)	Cultivated ha (4) Between brackets (4/3)
		On settlement sites				
1906	3104	1827	4931	3821	8227 (1.7)	2605 (0.31)
1910	5633	3477	9110	4454	7965 (0.9)	2793 (0.35)
1915	9692	3226	12,928	5318	12,877 (1.0)	4909 (0.38)
1920	11,449	3538	14,987	7545	11,865 (0.8)	6339 (0.53)
1925	14,177	3534	17,711	9153	13,284 (0.76)	8243 (0.62)
1930	16,561	2895	19,456	7524	10,580 (0.54)	8241 (0.78)
		Outside settlement sites				
1906	2954	9260	12,214	4402	89,906 (7.4)	7148 (0.08)
1910	8853	11,365	20,218	1973[a]	57,157 (2.8)	8270 (2.14)
1915	7569	12,489	20,058	5585	102,719[a] (3.5)	9118 (0.13)
1920	10,729	10,036	20,765	5164	53,959 (2.6)	16,348 (0.3)
1925	17,254	10,275	27,529	7754	79,476 (2.9)	16,909 (0.21)
1930	21,982	10,333	32,315	7308	119,689[a] (2.2)	13,163 (0.19)

Source: Adaptation of Colonial Reports, statistical appendices
[a]Probably a wrong number

average thus obtained for the number of plots in 1910 outside settlements is 5,035, and the number for the total acreage in the years 1915 and 1930 is 70,125. This is a conservative procedure because it assumes that the total acreage did not increase significantly despite population growth. Although the figures have 'a hectic gradient', my estimates may be considered a better approximation of the true state of affairs in the corrected years. For these reasons, the figures are no more than indicative.

Table 6.1 shows several trends over the first quarter of the twentieth century. The first observation is the sharp increase in immigrants, both on and off settlement sites. This increase was more than fivefold at the settlement sites and more than sevenfold outside the sites during the period under review. The number of non-immigrants hardly increased, which can be explained by the migration of Creole workers to other sectors. Nevertheless, the non-immigrant population at the settlement sites grew by about a thousand people during the period covered. Outside the settlement sites, the non-immigrant population grew by about the same number. The overall effect of these trends is that the population at the settlement sites quadrupled and that outside the settlement sites it increased by more than two and a half times. The increase was mainly accounted for by immigrants.

A second observation is that the population, as well as the available land, was smaller at settlement sites than outside them. This suggests that there was a link between settlement and plot size. Fokken (2018, p. 217) believes that the choice of plots outside settlement sites was mainly motivated by the low rent for uncultivated land compared to plots at settlement sites. Also, the small size of the plots at the settlement sites was less attractive. Another reason for settling down outside the settlement sites was the desire to escape the plantation regime.

But since the number of plantations was decreasing at an accelerated rate, this explanation is unlikely. It is more plausible that the plots at the settlement sites were too small to make a good living. In 1921, Leys (1921, p. 33) characterised settlement sites as follows: 'They are arranged in such a way that all resettled people are doomed to remain small. Nobody can get property, only a few rent a little more than two hectares of land … Nowhere can proper cattle breeding be carried out due to lack of pasture. Fertilising the depleted land is out of the question. There are no means of transport, the roads are poor and the construction of the land and buildings on it is injudicious. Doomed to perpetual poverty, those are the inhabitants … And yet agriculture in the settlements still compares favourably with small-scale agriculture outside.'

Leys (1921, p. 33ff) then described how hard the work on the plots of land outside the settlements was: deforestation, clearing, planting and the like under very unfavourable conditions that ranged from ant and mosquito infestations to heavy rainfall. With difficulty, one man could prepare and maintain 1 ha of land ready for building. Leys wrote this in 1921 and he had the Creole peasant community in mind. This was no coincidence. Table 6.1 shows that in the period before 1920, non-migrants outside settlement sites were in the majority. The plots of land outside the settlement sites were larger than those at the settlement sites, and on average—i.e.

as a ratio of area:population—the availability of land was also greater there (see the figures in brackets in the column 'total area' in Table 6.1).

Owning large tracts of land did not yet lead to higher yields as parts of the plots (both on and outside the settlements) remained undeveloped. On the settlement sites, the cultivated area increased from 31 to 78% in the period 1906–1930, and outside the settlement sites these percentages varied from 1 to 2%. So land was in abundant supply. The question arises as to why large areas of available land remained fallow. An obvious explanation is that the peasants were limited in their labour capacity to work larger pieces of land. The most common tools were the tjap (hoe) and the pickaxe. Given this level of technology, labour output was determined by the size of the family. But as explained in Sect. 2.3, family formation was limited in the first quarter of the twentieth century.

A second possible explanation is that peasants used only a small portion of land to restore the natural productivity of a previously used portion. However, there is no indication in the literature that this took place on a significant scale. In discussions on smallholder agriculture, agronomists complained about the inexpert way peasants conserved the natural productivity of the land. Here, most attention was focused on peasants' labour input, the need to replace human energy with animal energy, the use of machinery, the choice of crops grown, the introduction of new varieties and the improvement of planting methods (Beck 1924b, pp. 5–7; Hewitt 1922, pp. 14–17; Leys 1916a, pp. 5–11, b, pp. 1–8; Stahel 1933, pp. 11–29). These topics indicate the increasing focus of agronomists on the low labour productivity of peasants.

During World War I, food imports came to a halt, which allowed peasants—not just rice producers but all categories of food producers—to expand their production. This expansion is reflected in the increase in cultivated area, although not all the land was planted with rice (Heilbron 1982, p. VII). The increase in cultivated area stagnated, especially outside the settlement sites. In the period 1925–1930, it even experienced a decline. However, the agricultural population increased. This was due to the Great Depression, which drove many urban and other unemployed, predominantly Creoles, into the agricultural sector (van Lier 1977, p. 272ff). The steady increase in rice production cannot be explained entirely by an influx of people who temporarily sought refuge in agriculture, however. Rather, the growth in rice production was the result of the commissioning of new polders in Nickerie, the increasing use of light tractors and, above all, the use of new cultivation techniques and rice varieties (van Dijk 1940).

In the first decade of the twentieth century, the government tried to increase labour productivity by improving knowledge, cultivation practices and labour instruments. For example, McNeill and Lal (1915, p. 173) reported that '[t]he free Indian in this Colony is given all possible facilities for acquiring land of which enormous areas await settlement. The Government recognises that where virgin land or land long fallen out of cultivation is to be allotted to settlers the construction of drains, embankments, etc, must be undertaken by the government and not by the settlers'. And the report for the years 1924–1927 of the Department of Agricultural Experimental Station (DPL) in Suriname states (on pp. 31–35) that the government offered credits and land, introduced pumps to use fresh water from canals for irrigation and brought in iron ploughs from Indonesia because agriculturists were using home-made primitive implements that loosened the soil too superficially.

The annual report for the years 1931–1932 (DPL 1931, pp. 50–51) again mentioned ploughs being brought from the Dutch East Indies, that experiments with rice varieties from the Dutch East Indies were carried out and demonstrations were given on how to plough fields with oxen and then plant them. The department also provided seeds where peasants had to repay one-and-a-half times the quantity borrowed. 'The peasants, who do not yet have plough oxen, should therefore start planting the rice in rows, so that it can be weeded with a simple weeding device called a "landak". This device, which consists of a wooden roller with iron pins, not only weeds the soil, but also tills it lightly. As it turned out, this increases the yield' (DPL 1931–1932, pp. 62–63). Yields increased when weeding was followed by tillage and when tillage was replaced by the landak. Ploughs and oxen were sold by the department; sometimes the department also provided credit, sometimes it was itself a supplier (DPL 1931–1932, pp. 64–65).

6.5 Regional Differences

In a series of lectures on Surinamese agriculture published in 1924, Beck described the British Indian peasant as being distinguished by two characteristics: intensive farming, which he described as '[h]e can work the same piece of land year in, year out, which is only made possible to him by having manure at his disposal by keeping cattle'; and by 'his great thrift, which could turn to miserliness. For the sake of money, he personally grants himself even less than he actually needs for life' (Beck 1924b, p. 20). With this, he differentiated small farming according to ethnic backgrounds without considering the agrarian capital and specific circumstances of the disaffected peasants in more detail.

Nevertheless, Beck pointed to three geographically dispersed forms of agricultural enterprises among British Indians. The main concentration consisted of peasants who had settled around Paramaribo since the early twentieth century, supplying the city with rice, milk, cattle and vegetables. The latter two in particular provided year-round sales and income. The second group consisted of peasants in the adjacent Commewijne district who could sell their produce in the major towns, especially Mariënburg and Moengo. So prosperous were these peasants that some of them had cars with which they made money through transport services. The third group consisted of peasants in Saramacca and Nickerie, who depended mainly on rice production. The market in Saramacca was small, so quickly oversupplied, and a journey from Lower Saramacca to Paramaribo took 2–3 days. The high cost of transport and the unfavourable health situation in this area made attempting to make a living in Saramacca trying. In Nickerie, there was considerably more (polder) land available, but it was vulnerable due to the rainfall. And there, too, there was a serious marketing problem (Beck 1924b, pp. 21–23).

Annex 6.1 shows the indices of smallholder agricultural production, divided by settlements and beyond. The beginning of the indices (= 100) is the average production from 1905 to 1910. The explanatory notes show that the base of the index consists of different quantities. The base index for rice at the settlement sites is 3.8 million kg, while for outside the settlement sites it is 13.4 million kg, more than three times as much. Other big differences are cocoa (0.9 and 4.1 million kg, respectively), maize (0.3 and 5.4 million kg, respectively) and cowpeas (0.09 and 5.8 million units, respectively). Conversely, production at the settlement sites exceeds that of agriculture outside the settlement sites in terms of nuts and plantains. The other products do not differ much.

The indices can be read as the growth rate of production. Rice grew fastest at the settlement sites and matched the volume of production outside these sites. Coffee, plantains and coconuts also grew spectacularly, but mainly on the settlement sites. Apart from these increases, the production of small agriculture was stagnant in the first quarter of the twentieth century. It is the development of rice in particular that stands out, providing a sharp contrast to the feeling in the colony that this product had little future. It is generally said of rice that it is a culture of paupers and that rice peasants can never achieve prosperity, because the monetary value of the product per unit area is small. That proposition is arguable (Liems 1926, p. 376; see also De Klerk 1941, p. 104).

The growth of rice was regionally uneven, which was not surprising since the production factors also differed from district to district. The figures in Table 6.2 show the moving three-year average (to smooth out large fluctuations) growth of rice production.

The districts of Suriname and Nickerie were the main rice districts, followed by Commewijne and Saramacca. In 1931, Commewijne (especially the settlements of Hecht and Sterk, Kroonenburg and Johan and Margaretha) produced less than half of the rice produced by the district of Suriname (especially in the south of Paramaribo, Livorno, along the Pad van Wanica). Saramacca's share was a quarter

Table 6.2 Moving three-year averages of rice production in Suriname's rice districts, 1904–1931, in 1000 kg

Year	Nickerie	Coronie	Saramacca	Suriname	Commewijne
1904	30	–	166	537	89
1907	181	2	295	369	221
1910	289	6	320	1138	297
1913	329	4	319	1780	309
1916	787	9	940	2119	1029
1919	978	62	–	3911	3556
1922	2604	104	941	5223	3139
1925	5901	320	1320	6575	2866
1928	5588	821	1309	7456	2901
1931	7043	1016	2217	8623	4100

Source: Stahel (1933, p. 7)

of that of the district of Suriname. While rice production in the district of Suriname grew steadily, that in Nickerie increased in leaps and bounds. This can be understood from the commissioning of new polders: Van Drimmelen, Sawmill and Van Boonackerpolder, later the Corantijnpolder, and again a few years later the Henarpolder, Paradise, Margarethenburgpolder and the Hamptoncourtpolder (Stahel 1933, p. 9; see also Annex 5.3).

Bringing these lands under cultivation required great collective efforts, as shown in Photo 6.3. According to Stahel (1933, p. 6), the growth in production was partly related to the closure of some large farms and the release of labourers dedicated to rice cultivation, and partly to the persistently low rice prices since 1929, which peasants tried to compensate for by expanding their production. Proponents of middle-class agriculture pointed to the high degree of self-employment of Nickerian peasants, who generally took out higher loans from the peasants' loan banks (see Table 7.2). These peasants were in favourable natural conditions (land and water), were supported by the government and brought new polders into operation. The cooperative established around Sawmill Creek as early as 1905 was later a model for cooperative farming in other districts (Memorie 1919).

Although various authors pointed to regional differences, with the exception of Leys (1921) they hardly mentioned peasant poverty. However, this feature of peasant life can be derived from the then-prevalent view that rice farming was a form of subsistence for paupers (De Klerk 1941, p. 104; Liems 1926, p. 376). The poor health of the British Indians, as described by Lampe (1927) and De Cock Buning (1926), may also be underlined. Land was available but an insufficient indicator to express welfare levels. These differences in income and the (in)ability of households to make ends meet can already be deduced from the design of colonisation

Photo 6.3 First stone laying of the sluice in the Corantijnpolder, 1920

that forced peasants to work outside the home, while agricultural employment hardly existed.

6.6 Occupational Differentiation

Peasants were not only peasants but often practised other crafts as well. De Klerk (1953, pp. 111–112) listed the declared occupational backgrounds of landed immigrants who arrived on the ships *Howrah* and *Zanzibar*: soldier, shepherd, shopkeeper, busker, carpenter, mud seller, weaver, blacksmith, street robber, bricklayer, baker, woodcutter, opium preparer, schoolmaster, tailor, barber, writer and cook. The list is a selection to make clear that the immigrants had knowledge of crafts and other professions in addition to agriculture. Their declarations need not be completely correct—for instance, because the person provided incorrect data, because they had several professions of which only one was mentioned, or because the immigrants took certain work for granted or did not have a name for it.

These 'latent occupations' had already been practised during their stay on the plantations. Practical matters such as cutting hair, making or adjusting clothes, cooking and washing and carpentry, in addition to running shops and trading, providing credit, helping to raise cattle, growing vegetables and other services, were common (Choenni 2016; Fokken 2018). The peasantry could not be self-sufficient because numerous necessities such as salt, textiles, fuel and hardware had to be sourced from abroad as Suriname did not produce any of these goods itself. For these goods, the peasants depended not only on the border towns but on Paramaribo in particular, which was the most important import port. And more importantly, they needed to have money. Sale of produce was therefore decisive for their livelihood.

Professional differentiation on the plantations developed further during the period in which the peasants built their livelihoods. Two forms of differentiation can be distinguished. The first was derived from the peasants' prosperity and had a direct effect on agricultural production. On this, De Cock Buning (1926, pp. 42–43) writes: 'Insofar as these products are also eligible for export or for processing in husking mills (rice), purchases are made by British Indians, Chinese and Portuguese, who … often press the product to low prices by providing advances under harvest conditions. The exporters themselves also have their buyers, with whom a fixed price is contracted and who seek their profit from buying below that price. The small peasant usually gets a minimum price and is too little of a merchant to benefit from a favourable economic situation. In order to support him, government auctions were set up, but they had moderate success, as the British Indian buyers often set a maximum price …' Even if there were sales, the proceeds did not fall entirely to the peasants.

The second form consisted of practising old and new professions. As early as 1908, missionary Wenzel (1908, p. 6) wrote: 'In the vicinity of Paramaribo, numerous coolies have settled down. Mostly they grow vegetables and keep cattle. The produce of the land and milk they sell in the city.' The location of Paramaribo was not only related to the market. Fokken (2018, p. 219) points to specific factors that made this possible: '… the popularity and success of a particular government settlement was primarily related to the state of drainage and access to Paramaribo. Pad van Wanica, Kroonenburg, Livorno, and Saramaccapolder were well connected to Paramaribo, the latter three through waterways, while Pad van Wanica was located next to the railway which was opened in March 1905 … Immigrants generally used the social network they had established while living on the plantation to find out about post-indenture opportunities.'

The new occupational groups consisted of successful rice peasants and paddy mill owners. Van Traa (1946, p. 109), President of the Surinaamsche Bank (which acted as a central bank), observed about the years around the First World War: 'These were years of great prosperity for the small peasant and the British Indians in particular, with their industrious and thrifty, almost miserly nature, benefited greatly. Many of them became wealthy during those years and accumulated the capital that enabled them to engage in trade and commerce, and to give their children a good education, sometimes even to be able to study. In this way, the original British Indian immigrant-working class has grown into a fairly large middle class, which plays an increasingly important role in the life of Suriname, since entrepreneurs in trade and business as well as government officials, members of the liberal professions and commercial employees emerge out of them.' In addition to this differentiation within the group of rice peasants, De Klerk (1941, p. 105) points to the revival of old trades and the emergence of new ones, such as tailors, goldsmiths, car renters and tradesmen.

This occupational differentiation was already present on the plantations and continued among the peasants. It is important to distinguish two forms. In the first, several professions persisted in the agrarian milieu. Think of the rice miller, the barber, the shopkeeper, the pandit and the goldsmith. In another form of differentiation, professions outside agriculture emerged. In the first decades of the twentieth century, common occupations were those of milkman, charcoal seller and transporter (Beck 1924c; van Lier 1977). Entrepreneurship among Hindostanis, as pointed out by the aforementioned authors and later by De Klerk (1953), and even later by Gowricharn (1991), therefore arose mainly outside agriculture. Agriculture itself offered few opportunities for this.

6.7 Conclusions

This chapter has addressed the homemaking of the Girmitiya peasantry. This took three related forms: a cultural, a geographical and an economic one. The first consisted of the formation of an ethnic group, which enabled the mobilisation of ethnically specific knowledge about crops, plants, seeds and animals, as well as the labour of ethnic group peers. The second consisted of shaping the surrounding nature, especially deforestation and the construction of paddy fields, as well as the expansion and cultivation of vegetation and animal husbandry. The third was an outcome of the preceding two forms: the economic living enabled by the cultural and geographical forms. All homemaking was reflected in the landscape characterised until the late twentieth century by peasant farms consisting of rice fields, fruit trees and vegetable cultivation, as well as cows, chickens, ducks, goats and sheep around the house. Peasant farming shaped the cultural and geographical setting.

Homemaking by peasants was possible only when they had access to the land. Such access, according to most authors, would be the most important reason to settle in Suriname. A closer look revealed, however, that the Girmitiyas remained in Suriname for different reasons. The idea that they chose to stay in Suriname out of a passion for money and landholding or that they made that choice consciously is unconvincing. For that, there were too many, and sometimes compelling, circumstances that prevented the workers from returning. Another characteristic of the peasants was their limited labour capacity to cultivate the available land. The availability of land alone rarely indicates the prosperity of peasants.

With the outbreak of World War I, food imports declined and there were shifts in consumption from ground fruits to rice. The increase in rice production was driven not so much by increased labour productivity but by the opening up of new areas, especially in the Nickerie district where the land was also more productive. Across and within districts, rice production was probably uneven because production factors were distributed differently. This applied to almost all facets of the production process, in particular the size and quality of the land, the labour capacity as reflected in the presence or absence of family formation, the young age at which some peasants settled, the unequal knowledge of agriculture, the (im)possibility of investment from savings or access to credit, the transport possibilities and the location in relation to sales markets.

Remarkably, the government hardly paid any attention to the unequal availability of these production factors. It was not the people that were important but the production, and in particular the domestic food supply that could also reduce or even replace food imports, and any surplus that could be exported. From almost all reports, a focus bordering on obsession emerges with the plantations and later with surplus production by the smallholders. Especially after World War I, the focus was on the relatively strong peasants in the Nickerie district. It was a betting on the strong that would prove disastrous in the long run.

The unequal disposition of production factors not only makes it clear that the Girmitiya peasants formed an economically fragmented community but it also suggests that the chains of production, from homemaking to the cultivation of rice and other crops, husking, transport and sales, developed unevenly. These required activities were compelling and mostly developed in and around Paramaribo (i.e. including the district of Suriname), and to a lesser extent in Commewijne and Nickerie. In Nickerie, in particular, the problem was that agricultural production could boom, but sales were limited, an issue that was also typical for peasants in the Saramacca district. As a consequence, large numbers of peasants lived in abject poverty because production was inadequate, or because they failed to sell produce and supplementary employment in agriculture had shrivelled. Although most agronomists were optimistic about the ability of British Indians to survive, eyewitnesses of the time emphasised the poverty in which they lived. The homemaking was a poor home for most Girmitiya peasants.

References

Adhin JH (1960) Over de 'joint family' der Hindostanen. Nieuwe West-Indische Gids 40(1):17–27

Beck S (1924a) De Verschillende Bedrijfsvormen in den Klein–Landbouw met Betrekking tot Rasaanleg en Volksgewoonten. In: Beck S (ed) Lezingen over Surinaamsche Problemen. Kersten, Paramaribo, pp 1–31

Beck S (1924b) De Creolen en de Klein–Landbouw. In: Beck S (ed) Lezingen over Surinaamsche Problemen. Kersten, Paramaribo, pp 1–34

Beck S (1924c) Landbouwbedrijfsvorm en landbouwopleiding. In: Beck S (ed) Surinaamsche Problemen. Kersten, Paramaribo, pp 1–44

Bhagwanbali R (2010) Hindoestaanse migranten onder het indentured labour systeem naar Suriname, 1873–1916. De nieuwe awatar van slavernij. Amrit, Den Haag

Bisnauth D (2000) The settlement of Indians in Guyana 1890–1930. Peepal Tree Press, Leeds

Blunt A (2003) Collective memory and productive nostalgia: Anglo-Indian home-making at McCluskieganj. Environment and Planning D: Society and Space 21:717–738

Brandon HG (1925) Het Surinamevraagstuk. West-Indische Gids 6(7):481–560

Braverman H (1974) Labor and monopoly capital: The degradation of work in the twentieth century. Monthly Review Press, New York

Burger JW (1928) Vergelijkend Overzicht van de Immigratie en Blijvende Vestiging van Javanen en Brits-Indiërs in Suriname. De Economist 77(1):422–445

Carrier J (2018) Moral economy. 'What's in a name? Anthropological Theory 18(1):18–35

Choenni C (2016) Hindostaanse contractarbeiders 1873–1920. Van India naar de plantages in Suriname. LM Publishers, Volendam

Comins DWD (1892) Note on emigration from the East Indies to Surinam or Dutch Guyana. Printed at the Bengal Secretariat Press, Calcutta

De Cock Buning W (1926) Rapport over Suriname, uitgebracht aan de Vereeniging van Kamers van Koophandel en Fabrieken in Nederland

De Klerk CJM (1941) De Brits-Indiërs in Suriname. Nieuwe West-Indische Gids 24(1):97–117

De Klerk CJM (1953) De Immigratie der Hindostanen. Urbi et Orbi, Amsterdam

Easthope H (2004) A place called home. Hous Theory Soc 21(3):128–138

Emmer P (1984) Een goede keus? De migratie van vrouwelijke contractarbeiders van Brits-Indië naar Suriname. OSO, Tijdschrift voor Surinaamse Taalkunde, Letterkunde en Geschiedenis 3(2):213–234

Emmer P (1985a) De werving van kontraktarbeiders in India voor tewerkstelling in Suriname, 1872–1916. OSO, Tijdschrift voor Surinaamse Taalkunde, Letterkunde en Geschiedenis 4(2):147–157

Emmer P (1985b) The great escape: The migration of female indentured servants from British India to Surinam, 1873–1916. In: Richardson D (ed) Abolition and its aftermath: The historical context, 1790–1916. Frank Cass, London, pp 245–266

Fokken M (2018) Beyond being koelies and kantráki. Constructing Hindostani identities in Suriname in the era of indenture 1873–1921. Doctoral dissertation, University of Groningen, Groningen

Geertz C (1983/2000) Local knowledge: Further essays in interpretative anthropology. Basic Books, New York

Goerdayal L (2019) Planten met spirituele en devotionele betekenis voor Hindoes in Suriname en Nederland. In: Baldewsingh R, Boedhoe N, Mahabier B (eds) Sarnámi Sanskirti 2. Enkele facetten van de Hindustaanse geschiedenis en cultuur in Suriname en Nederland. Uitgeverij Surinen, Den Haag, pp 243–264

Gowricharn RS (1991) Economische transformatie en de staat. Over agrarische modernisering en economische ontwikkeling in Suriname, 1930–1960. Doctoral dissertation, University of Utrecht

Gowricharn RS (2013) Ethnogenesis: The case of British Indians in the Caribbean. Comp Stud Soc Hist 55(2):388–418

Gowricharn RS (2016) Hindostaanse heritage in Suriname. OSO, tijdschrift voor Surinaamse Letterkunde en geschiedenis, November, 115–129

Gowricharn RS (2020) Multiple homemaking. The ethnic condition in the Indian diaspora. Routledge, London

Gowricharn RS (2023) Beyond the Kala Pani. Indenture Papers Studies on Girmitiyas 3:25–38. https://girmit.ac.fj/wp-content/uploads/ip-2023/gowricharn.pdf

Grierson GA (1885/1975) Bihar peasant life, being a discursive catalogue of the surroundings of the people of that province, with many illustrations from photographs taken by the author. Cosmo Publications, Delhi

Haraksingh K (1999) Indentured and self-emancipation. In: Lamming G (ed) Enterprises of the Indies. Institute of the West Indies, Tunapuna, Trinidad & Tobago, pp 38–42

Hassankhan M (1993) De Immigratie en haar gevolgen voor de Surinaamse samenleving. In: Gobardhan-Rambocus L, Hassankhan M (eds) Immigratie en Ontwikkeling. Emancipatie van Contractanten. Anton de Kom University, Paramaribo, pp 11–35

Heilbron W (1982) Kleine boeren in de schaduw van de plantage. De politieke ekonomie van de na-slavernijperiode in Suriname. Doctoral dissertation, Erasmus University Rotterdam

Hewitt WE (1922) Nickerie als landbouwdistrict. West-Indië, Landbouwkundig Tijdschrift voor Suriname en Curaçao 7(4):5–20

Hoefte R (1998) In place of slavery. A social history of British Indian and Javanese laborers in Suriname. University Press of Florida, Gainesville

Kraijo M (2022) Destined to leave Hindustan for Suriname? Explaining the motivation behind repatriation and settlement of Hindustani labour migrants in Suriname, 1873-1940. Tijdschrift voor Sociaal-Economische Geschiedenis (TSEG) 9(3):37–68. https://doi.org/10.52024/tseg.10894

Kruijer G (1949) Landbouw-Cooperatie onder de Aziatische bevolkingsgroepen van Suriname en Brits-Guyana. Nieuwe West-Indische Gids 30(1):209–235

Kumar A (2017) Coolies of the Empire. Indentured Indians in the sugar colonies 1830–1920. Cambridge University Press, Cambridge

Lal B (1983) Girmitiyas. The origins of the Fiji Indians. Australian National University Press, Canberra

Lal B (2000) Chalo Jahaji. On a journey through indenture in Fiji. Australian National University Press, Canberra

Lampe PHJ (1927) Suriname. Sociaal–Hygiënische Beschouwingen. De Bussy, Amsterdam

Laurence K (1994) A question of labour: Indentured immigration into Trinidad and British Guiana 1875–1917. Ian Randle Publishers, Kingston

Leys JJ (1916a) Het bemesten van onze rijstvelden. West-Indië, Landbouwkundig Tijdschrift voor Suriname 1(1):1–11

Leys JJ (1916b) Het bemesten van onze rijstvelden. West-Indië, Landbouwkundig Tijdschrift voor Suriname 1(2):1–8

Leys JJ (1921) De Kleine Landbouwers en het Klein Landbouwbedrijf. West-Indië, Landbouwkundig Tijdschrift voor Suriname en Curaçao 6(1):31–44

Leys JJ (1925) De veeteelt in Suriname. Nieuwe West-Indische Gids 6(1):405–418

Liems JA (1918) Is de Creool voorbestemd om dienstbaar te worden aan den Brits–Indiërs? West-Indië, Landbouwkundig Tijdschrift voor Suriname en Curaçao 3(3):140–144

Liems JA (1926) Over den kleinen landbouw in Suriname. West-Indische Gids 7(1):371–384

Mahabir K (2004) Animals from India in the Caribbean. Zoological survey of India 103(Parts 1–2). Government of India, Kolkata, p 179–183
Mahabir K (2008) Medicinal and edible plants used by East Indians of Trinidad & Tobago. Chakra Publishing House, Madison
Mayer A (1966) Quasi-groups in the study of complex societies. In: Banton M (ed) The social anthropology of complex societies. Tavistock Publications, London, pp 97–122
McNeill J, Lal C (1915) Report to the government of India on the conditions of Indian immigrants in four British colonies and Surinam, Part II. London
Memorie (1919) Memorie van het Comité tot Bevordering van den Middenstandslandbouw in Suriname, op 29 augustus 1919 den Gouverneur Aangeboden. West-Indische Gids 1(2):163–176
Mintz S (1983) Reflections on Caribbean peasantries. New West Indian guide 57(1–2):1–17
Nayar P (2015) The postcolonial studies dictionary. Wiley Blackwell, Oxford
Raghoenandan U (1994) Etnobotanisch onderzoek bij de Hindostaanse bevolkingsgroep in Suriname. UvS/BBS (Nationaal Herbarium van Suriname)
Ramdin R (2000) Arising from bondage: A history of Indo-Caribbean people. Tauris Publishers, London/ New York
Samaroo B (2017) Changing Caribbean geographies: Connections in flora, fauna and patterns of settlement from Indian inheritances. Unpublished manuscript, History Department, University of the West Indies, St. Augustine, Trinidad
Stahel G (1933) De Rijstcultuur in Suriname; Amsterdam, z.j. (overdruk uit De Indische Mercuur, 22 en 29 november, en 6 dec 1933), publication of the Landbouwproefstation te Paramaribo, no. 2
Stiglitz J, Walsh C (2006) Principles of Micro-Economics. Norton & Company, New York/London
Van Andel T, Ruysschaert S (2014) Medicinale en rituele planten van Suriname. LM Publishers/ VACO, Arnhem/Paramaribo
Van Dijk HN (1940) Mechanische Rijstcultuur: Economische en Technische Resultaten Na Zes Jaar Pioniersarbeid te Nieuw–Nickerie, Suriname. Amsterdam: Paris
Van Dusseldorp DBWM (1963) Het Centraplan voor Suriname: Some aspects of town and country planning in a developing country. Wageningen, Veenman & Zonen
Van Lier R (1977) Samenleving in een grensgebied: Een sociaal–historische studie van Suriname. Emmering, Amsterdam
Van Traa A (1946) Suriname 1900–1940. Van Hoeve, Deventer
Van Wouw JJ (1949) Het Departement van Landbouw-Economische Zaken in Suriname in 1945. Nieuwe West-Indische Gids 30(1):244–249
Wenzel T (1908) Schetsen uit den Zendingsarbeid der Evangelische Broedergemeente onder de Brits–Indiërs in Suriname. Vonk & Co

Chapter 7
The Politics of Ruralisation, 1930–1950

Abstract This chapter examines the migration of Girmitiya peasants to the city and their orientation towards entrepreneurship and white-collar occupations. The government tried in vain to reverse the urbanisation by pursuing a ruralisation policy. The analysis is outlined as follows. Section 7.2 addresses the prevailing agricultural issues at the time. These consisted mainly of poor ethnic cohesion, which Governor Kielstra sought to solve by pursuing a ruralisation policy. Section 7.3 discusses production factors and the problems that were involved, specifically their productivity. Section 7.4 highlights that agricultural development was regionally uneven and rice production became concentrated in certain districts, among which Nickerie retained its advantageous position. Section 7.5 raises the issues of processing and marketing, problems that inhibited the growth of rice production. These represent a major concern for the colonial administration as it hoped that small-scale agriculture, specifically rice, would become an alternative to plantation agriculture. Section 7.6 looks at peasant poverty, a little-discussed social issue that has taken different forms geographically.

Keywords Poverty · Ruralisation · Uneven development · Agrarian productivity · Nickerie

7.1 Introduction

In the period 1895–1920, British Indian indentured labourers were transformed into peasants in Suriname. This development was stimulated by intensive government education and demonstrations aimed at increasing the efficiency of the production process. Although rice production rose, the surplus to export was not significant. Agricultural productivity, infrastructure, maintenance and management were considered the main problems preventing a take-off of rice production. While the agriculturalists of the time were mainly concerned with the business aspects of farming, they continued to interpret communities in ethnic terms, i.e. in terms of Hindostani,

R. Gowricharn, *The Girmitiya Peasants in Suriname*,
https://doi.org/10.1007/978-3-031-67961-2_7

Creole, Javanese and Dutch peasants. This was the case until after World War II (De Vries 1965).

The entanglement of culture and economics was acknowledged at that time. In the Netherlands, colonial state economics had developed, a forerunner of development economics, of which professors Kielstra (1918) and Boeke (1930), respectively, held the first chairs at the universities of Wageningen and Leiden. It was also a respectable tradition in agricultural economics at the time to link social, geographical, biological, economic and political forces. But this synthesis was lacking among most agronomists in Suriname. They were mainly, if not exclusively, focused on the farm level to increase agricultural production and had little regard for the binding and cooperative forces of the peasants. When Kielstra was appointed governor of Suriname in 1933, he tried to strengthen this neglected connection between culture and economy. This was his major contribution to the development of Hindostani peasantry.

However, despite his experience and benevolence, Kielstra overlooked two major issues. The first was that ethnic community development and the modernisation of agriculture will inevitably affect the communal capital of the Girmitiyas. The modernisation rendered many peasant households independent as it made the traditional cooperation superfluous. Second, as productive forces were unequally distributed, poverty was widespread among the Girmitiya peasants. As a result, many peasants were not in a position to use the available resources as any use entailed some kind of investment.

This chapter examines the developments of the Hindostani peasantry in the period 1930–1950. The long-standing poverty among Hindostani peasants continued in the period up until after World War II. That encouraged migration to the city and an orientation towards entrepreneurship and white-collar occupations. The government tried in vain to reverse the urbanisation by pursuing a ruralisation policy. First, at the macro level, the migration from agriculture to urban areas diminished the size of the farming population. This flight was especially manifested shortly before and during World War II when more and more peasants, artisans and young people moved to the city for work and school. Second, at the community level it led to a decline in ethnic cooperation. The households improved their production process and required less assistance from kith and kin. The ethnic ties remained but played a declining economic role in agriculture.

The next section addresses the prevailing agricultural issues at the time. These consisted mainly of poor ethnic cohesion, which Governor Kielstra sought to solve by pursuing a ruralisation policy. Section 7.3 discusses production factors and the problems that were involved, specifically their productivity. Section 7.4 builds on this and highlights the fact that agricultural development was regionally uneven. As a result, rice production became concentrated in certain districts, among which Nickerie retained its advantageous position. Section 7.5 raises the issues of processing and marketing, problems that inhibited the growth of rice production. These represent a major concern for the colonial administration as it hoped that small-scale agriculture, specifically rice, would become an alternative for plantation agriculture. Section 7.6 looks at peasant poverty, a little-discussed social issue that had taken different forms geographically. The concluding section summarises and evaluates ruralisation and agricultural development up to the 1950s.

7.2 Agrarian Issues

Agricultural policy after the Depression years was to some extent a continuation of the promotion of smallholder agriculture. However, around 1920 the Netherlands was fed up with Suriname. All previous attempts to pull the colony out of the doldrums, which, incidentally, were all half-hearted according to Kielstra (1927), had failed. The colony was characterised as 'the sick man', where 'malaise prevailed in permanence' and where everything failed. Many a Dutch minister acknowledged not knowing what to do with the country (Ramsoedh 1990, p. 5). However, there was also criticism of colonial policy. This mainly centred on the austerity measures, covering deficits with subsidies instead of loans and a lack of courage to tackle problems on a grand scale. At the beginning of the 1930s, Suriname experienced several workers' uprisings that also highlighted the political dimension of the Suriname question (Ramsoedh 1990, pp. 39–41). The call for a 'strong man' to rule the colony with an iron hand and vision became louder. It was in this context that Professor J.C. Kielstra was appointed governor of Suriname in 1933.

Kielstra's ten-year rule had important and widespread political consequences for Suriname. In particular, his efforts to strengthen the position of small peasants was important. Kielstra had had a career as a senior colonial official in the Dutch East Indies and later as a professor of colonial economics at the Agricultural University in Wageningen. Before coming to Suriname, he had dealt extensively with colonial issues. Those were analysed by Dutch scholars mainly from the experience gained in the Dutch East Indies (Aas 1980, pp. 221–224) (Photo 7.1).

This is not the place to present Kielstra's views at length (Kielstra 1918, 1927, 1946; Ramsoedh 1990, pp. 48–56). What is important is that he had regretted that Dutch capital had failed in Suriname. In his view, plantation agriculture could still be saved, but that would require large investments in the transmigration of Javanese workers (Kielstra 1927, p. 195; Ramsoedh 1990, p. 117). Presumably for these reasons, he sided with the small peasants in Suriname. That choice was a clear break in Suriname's agrarian policy in which the interests of the planters were central. This did not mean that Kielstra disregarded the planters' interests, but he had devoted himself more than his predecessors to the problems of the small peasants.

One of the issues Kielstra had had to deal with since his arrival in 1933 was geographical mobility and underutilisation of land and infrastructure by peasants. These increasingly sought work outside agriculture, which led to urbanisation, among other things. This problem was exacerbated in the 1930s by increased unemployment due to the decline of export sectors (sugar, cocoa, gold and rubber), and by the return of Surinamese workers from the Netherlands Antilles. In the face of growing urbanisation, Kielstra developed a 'ruralisation policy' through which he aimed to promote communities based on ethnic-religious principles. 'The lack of a social

Photo 7.1 Prof. J.C. Kielstra, governor of Suriname, 1942

structure is a serious obstacle to the development of good indigenous agriculture. It is the Government's duty to create and strengthen such a social organisation' (Kielstra 1946, p. 353). Kielstra hoped to counter urbanisation and that communities would take (more) care of infrastructure.

In official circles, there was the idea that by granting ownership, peasants would make better use of the land. In the decade after Kielstra took office, a discussion on this issue developed without any decision being made (Bosz and Albrecht 1954, pp. 275–288). One exception concerned the allotment of plots of land owned in the Nickerie district. This issue had 'not been an unqualified success and was therefore discontinued in 1949. On several occasions, the District Commissioner complained that he no longer had any control over the residents after the relinquishment of ownership, which is why he proposed leasing' (Bosz and Albrecht 1954, p. 274). Administrators believed that with the prevailing lease terms, an undisturbed, continuous and quiet possession was guaranteed that could also be transferred as an inheritance.

A third problem Kielstra tried to address was the strengthening of the social structure of the agricultural population. In 1934, 'civil servants were instructed to develop a sense of belonging in the countryside and to promote the founding of social and cultural associations' (Kruijer 1949, p. 224). In 1938, Kielstra instituted village ordinances to promote the collective interests of, and by, peasants. These measures were part of an attempt to improve production conditions and strengthen peasants' social cohesion. Particular focus was on: choosing areas that allowed uniform agriculture; improving transport, irrigation and drainage; providing loans;

clear legal titles to land; education and extension. In this context, water resources were investigated during his reign with a view to improving old agricultural areas and for the purpose of bringing new areas under cultivation (Junker 1940; de Vos 1938).

Also important for strengthening the farming community was the encouragement of cooperatives and the recognition of the cultural 'idiosyncrasies' of ethnic groups. Cooperatives were needed to promote collective interests. But this was a formal cooperation and not traditional cooperation as in neighbourly aid. The cooperation envisaged was focused on infrastructural facilities such as the tilling of land, the construction of roads and sluices and the installation of tubes (Kruijer 1949).

Ethnic differences in the production process were not a reason to change the colonisation pattern. 'A British Indian "joint family" may plough more land than a Negro or Javanese peasant in the West Indies, … but this does not necessarily imply a fundamental difference in the setup of the farm' (Kielstra 1946, p. 338). Fully in line with the prevailing thinking at the time, Kielstra also considered racial characteristics important, although he was less outspoken than his contemporaries (cf. Beck 1924). Javanese are not inclined to emigrate, according to Kielstra, and '[w]hat feels compelled to emigrate … are the more or less declassified elements in native society, not inclined to regular labour, accustomed to idleness and not appreciated in their environment ' (Kielstra 1927, p. 185).

In order to strengthen the cultural heritage, Kielstra gave legal force to the traditional full marriage of Hindus and Muslims. That had been requested by the British Indians since 1910 and had since been blocked by the Koloniale Staten (colonial parliament), in which light-skinned Creoles and Jews were over-represented (Adhin 1969; Ramsoedh 1990, p. 122ff). Kielstra also made an attempt to solidify the ethnic community by colonising Javanese families as peasants instead of employing them as indentured labourers. In 1930 and 1931, 270 farming families, totalling 977 people, were brought in for smallholder farming. In 1937, this endeavour was followed up by the establishment of three Javanese-style villages (dessa) in the Saramacca district. In 1939, another 383 Javanese families were brought in. It was a form of economic development through 'community development' consisting of improving production conditions, assigning self-responsibility and formalising it in village communities.

During Kielstra's reign, smallholder agricultural production increased significantly. Annex 7.1 shows rice production for the years 1929–1948. It reveals that production during that period was erratic but rising, peaking at around 50 million kg. It is also striking that only a small proportion of it was exported, although the export price over the whole period was higher than the domestic price for rice. No clear relationship can be observed between a higher price and the volume of rice exports. In the years with the highest export volume (1937–1939), the difference from the domestic price was small, while conversely, during the years with the highest export price (1944–1948), the volume of rice exported was at a lower level than that in the Depression years. Thus, price was not an explanatory factor of the increase or decrease in production.

Annex 7.2 shows that the area available fluctuated in the period 1931–1948, but that the portion devoted to rice doubled. Land under cultivation was expanded and a larger part of it was planted with rice. The total number of ha decreased from 120,410 to 64,234, the area of land under cultivation also decreased, albeit less (from 23,671 to 18,640 ha), while the rice area increased from 8132 to 19,465 ha. It should be noted that the number of ha earmarked for rice cultivation in 1948 was larger than the number of ha under cultivation. The trend that a larger proportion of cultivated land was devoted to rice cultivation is indisputable.

The mixed (rice) farm was not a capitalist enterprise where price developments determined the preference for crops grown and where production factors such as labour were mobile. If peasants became mobile, it was usually to leave agriculture. The Surinamese literature has frequently pointed out that both a price rise and a price fall could lead to an increase in production. The latter occurred mainly to compensate for an income drop (Schütz 1941, p. 92; Van Traa 1946, p. 109). Expanding production presupposed that the market was not saturated and that there was spare capacity (land, labour, means of production, including seeds and cows). Usually there was enough land, but peasants lacked the ability to cultivate it.

The reaction to falling and rising prices described above did not apply to every ethnic group. Javanese, in particular, produced hardly anything for the market but more for their own consumption. This made them more resilient in times of unfavourable price developments. For instance, De Vries mentions that the falling price of rice after 1920 did not affect their enthusiasm for settling on an agricultural plot. 'Rice prices were allowed to fall; Javanese peasants were not dependent on sales. They grew enough for themselves and, in the case of shortages, they could still call on their better-off relatives, friends and acquaintances, while the necessary cash income could also be obtained through wage labour on the plantations or, increasingly, from Hindostani rice planters' (De Vries 1965, p. 179). Thus, rice production depended not only on the availability of production factors but also on the ethnically varied efforts of peasants.

7.3 Production Factors

Rice production could increase by combining existing resources more efficiently. This more efficient combination was gauged by the increase in land yields, cultivation methods and yields of different rice varieties. In addition, agronomists disseminated agricultural knowledge, demonstrated the use of available resources and introduced new means of production. They also provided training and education on how to combat the damage caused by rats, birds, ants and insects, both in the field and in storage. Expansion of labour required either external help or an increase in labour productivity. The latter had changed traditional ethnic ties by the end of the 1930–1950 period. Help from peer groups was less required, as it could be replaced by borrowing from the Boerenleenbanken (Agricultural Loan Banks) and employing labour-saving inputs such as cows and donkeys.

7.3.1 Land

Expansion of rice acreage was difficult because of the suitability of land and because infrastructural facilities were unevenly distributed among the various districts. The 'infrastructural' facilities consisted of a system for draining excess water and for providing the land with water in periods of drought (Junker 1940, p. 232). This was done by building water canals or using natural channels and by constructing tubes and sluices. It was mainly the new agricultural areas, especially the polders, that needed these facilities, while peasants in the old areas used the inherited systems of the plantations. In addition, roads and bridges had to be provided or improved.

Expansion of the production area could occur when irrigation was put in place. Only in a few places (notably in Nickerie) was no soil preparation or irrigation needed because of the favourable natural location (Hewitt 1922). Expanding rice acreage was therefore not always possible, as the necessary infrastructure was usually lacking. Expansion was also limited by estate division, which led to the fragmentation of land and in turn required extra effort in terms of maintenance and supervision.

In the Commewijne district, where the plantation system prevailed, cocoa and coffee plantations were parcelled out and issued in plots with the planting. The crop assortment here was not only more mixed but it was also more stable over a longer period of time. The peasants were generally Creoles (Hewitt 1941). Since most of the fertile land was already in use by the plantations, in government schemes or on settlements outside them, the rectangular pieces of land could not be expanded. A peasant in this area who expanded their farm could get a plot located elsewhere, however. Farms could thus dispose of several plots. Over the period 1931–1948, the district of Suriname had the most and, on average, the largest plots. But the average cultivated area in Nickerie per plot was larger (Gowricharn 1991, Appendix B7). In other words, the size of plots in Nickerie was smaller on average than in the Suriname district, but in Nickerie on average a larger proportion of land was cultivated.

7.3.2 Population

According to Annex 7.3, during the period 1931–1948, the immigrant population in smallholder agriculture increased significantly in the Nickerie and Saramacca districts. In the district of Suriname, the immigrant population declined after rising untill 1942, while growth stagnated in the Commewijne district. The Nickerie and Suriname districts then had relatively large numbers of Hindostani peasants. Saramacca initially had a large number of Hindostani peasants whose share of the population declined with the increasing colonisation of Javanese in that district. Commewijne had a mixed Javanese-Hindostani-Creole population (Van Dusseldorp 1963, Annex 2). This mixed composition was also found in the district of Suriname, which was an 'urbanisation district': migrants who moved to the city often settled in and around the city of Paramaribo and were administratively classified as residents of the district of Suriname.

The ethnic composition of the district population had consequences for the volume of rice production. The districts with the largest concentrations of Hindostanis also produced the most rice (Annex 7.4). The districts of Saramacca and Commewijne possessed a small, though not negligible, share in the total rice production, while the predominantly Creole Marowijne district's share was negligible. In these districts, the absolute increase in available land was small while the percentage of land under cultivation also increased less rapidly and even decreased. In contrast, cultivated land increased sharply in the main rice districts, Suriname and Nickerie. Especially in the Nickerie district, production was gradually modernised, first by replacing human energy with animal energy and later by using tractors (Gowricharn 1991).

The departure of peasants from small farming can be partly explained by the fact that some found employment in other sectors. These include agricultural trade, the establishment of shops, specialisation in horticultural crops, transport, crafts and white-collar professions (Hewitt 1922; Memorie 1919; Schütz 1941).

As mentioned earlier, the peasantry was ethnically fragmented. The descendants of Dutch peasants were mainly engaged in cattle farming and the supply of milk and meat (Verkade-Cartier van Dissel 1937). These peasants owned a medium-sized farm; they were actually farmers who were traditionally familiar with raising cattle. Another group consisted of Javanese peasants who remained in the settlements and engaged in subsistence production (Van Wengen 1972). For a long time, their traditional method of rice harvesting involved the use of small knives (called 'ani-ani knives'), which rendered a low labour productivity. The Javanese did not expand rice production for commercial purposes. The Creole peasants had specialised in cocoa and coffee cultivation (especially in Saramacca) and in coconut cultivation, which they combined with pig farming (De Vries 1965). All groups owned their own agricultural human capital and were in very limited competition with each other.

7.3.3 *Technology*

As early as 1910, the government tried to replace the use of sticks, hewers and tjaps for rice cultivation with ploughs. But it wasn't until the second half of the 1920s that this tool was adopted on a significant scale. Adoption of the tool was slow due to the lack of draft cattle. 'The oxen needed for ploughing, which are only needed for ploughing for a short time, … have no work for most of the year. Only when draught animals have enough pasture available to take care of themselves when not in use does it pay off for the small peasant to buy ploughs and keep draught animals' (LPS 1924/1927, p. 16). But the size of the allocated land was not up to the task. Nevertheless, the government made an effort to modernise implements.

In 1925, the colonial government introduced six ploughs to be loaned to peasants. By 1927, there were 23 peasants with ploughs made in Suriname. Their use was conditional on having draft animals. Agricultural experts demonstrated the use of ploughs on specially created fields, and on these occasions information was also provided on seed selection, product mix, method of cultivation and so on. Harrowing was also improved by the introduction of specially designed plates by the Agricultural Experimental Station Suriname (LPS). These instruments enabled the peasants to cultivate a larger area. Sickles were used for harvesting and were mainly in vogue among Hindostanis. As a result, the gap in labour productivity between the different population groups increased (Photo 7.2).

The use of draft animals was a concern. The number of cattle in the colony was small and had to be maintained by imports. The increase in cattle numbers encountered problems because, apart from Hindostanis, who were predominantly Hindu, many peasants preferred to slaughter the animals to meet the increasing urban

Photo 7.2 Harrowing a paddy field with the peasant standing on a wooden structure pulled by animals, ca. 1925

demand for meat. 'Quite general is the phenomenon that after rice planting, plough oxen are sold to butchers because the owners are reluctant to keep these otherwise useful animals in stalls for 5 months (the time the rice is in the field) and feed them when they are not productive. Shortly before ploughing time, the peasants broach a new set of young animals for ploughing, which after ploughing also end up in the slaughterhouse' (LEZ 1941/1944, p. 54; for earlier periods, see Leys 1920a, pp. 69–75). As a result, there was a persistent scarcity of draft animals. The import of special breeding bulls, stock breeding and cattle feeding received increasing attention in the annual reports of the LPS. The government veterinarian also paid due attention to the care and diseases of these animals and meat inspection (Frickers 1939, pp. 28–46).

In the Nickerie district, the increase in plough oxen was striking as the following table shows.

Table 7.1 shows that in the 5 years from 1929, the number of plough oxen in the Nickerie district increased sharply, but the average number of oxen per rice peasant continued to fluctuate between 2.5 and 3. Annex 7.6 shows that beef cattle (cows and oxen), which can be a substitute for plough oxen, in the district of Suriname were more numerous than those in other districts and that their numbers grew most strongly in the district of Nickerie. However, the large cattle herd in the district of Suriname cannot be fully related to rice production since a proportion was accounted for by the Dutch peasants who kept cattle as dairy cattle and as slaughter cattle (Verkade-Cartier van Dissel 1937).

For small peasants, investments were needed to buy a number of cows, while the maintenance of the animals was limited by insufficient grazing land and not enough suitable cattle feed. The latter problem was particularly exacerbated in the rainy season as grass flooded (Snellen 1933, p. 148). Draft animals were important for rice production as they were used for ploughing, harrowing, threshing and transport, in addition to the delivery of milk and possibly meat. The expansion of livestock increased the demand for land and credit, and it also became a custom to hire out draft animals to peasants in the area. These services were paid in kind or in money. But as more and more peasants bought a number of cows, renting out these animals became redundant.

Table 7.1 Increase in plough oxen in Nickerie, 1929–1933

Year	Number of plough Oxen (1)	Number of peasants (2)	1:2
1929	10	4	2.5
1930	16	6	2.7
1931	57	20	2.9
1932	122	45	2.7
1933	138	50	2.8

Source: LEZ 1931/1932, p. 59

7.3.4 Credit

Peasants' demand for money stemmed from both consumptive and productive needs. Consumptive needs could be financing a child's marriage, building or repairing a house, repaying an old debt, buying durable household goods and so on. In this context, De Klerk (1941, p. 108) notes about Hindostanis: 'In certain customs, sanctioned by tradition, they have a real herd mentality, so that they don't dare deviate from them out of fear of people, e.g. throwing big parties at weddings, even though it would ruin them.' Productive spending included the purchase of land, draft animals, storage facilities and the payment of labourers.

Credit could be obtained from formal and informal capital markets. The formal market was small and consisted of the Colonial Postal Savings Bank, the Surinaamsche Bank and the Boerenleenbank. The latter banks were run on a cooperative basis and set up for smallholder peasants. Access to this source of credit depended heavily on the peasant's reputation and the intercession of the district commissioner. In fact, the other banks were inaccessible to small peasants as their clientele consisted mainly of the plantations and non-agricultural enterprises.

The informal capital market was more accessible, but very fragmented and extremely expensive. The lenders here consisted of different categories of self-employed people: doctors, lawyers/notaries, shopkeepers, rice mill owners, traders and rich peasants. Most peasants needed advances for harvesting, transportation and rice husking. These were provided at interest rates as high as six times the market rate and could range from 2 to 25% per month (LEZ 1920, p. 71, 1923, p. 77; Van de Walle 1945, pp. 68–69). In the Nickerie district, the purchase of land and draft animals, the construction of storage barns, the reclamation of land, etc. were mostly financed from the peasants' own savings or loans from the Boerenleenbank.

Table 7.2 shows the credit extended by the Boerenleenbank during the period 1931–1949. The Nickerie district had the highest demand for credit, followed by the Commewijne district. During WWII, the credit demand slumped and then experienced some outliers. The average loan amount was also highest in Nickerie. The increased production in that district was therefore accompanied by higher debts. Note that loans were not only used to expand production but also to absorb setbacks or replace means of production. In these cases, the investments need not be reflected in the expansion of acreage or increase in production.

The use of agricultural credits varied by district and ethnic group. In the old agricultural areas where production conditions were less favourable, credits were more likely to be used for consumption purposes. This was predominantly, though not exclusively, stated for Creoles, as most of them were reluctant to repay the credits (Liems 1918). In expanding rice districts, on the other hand, peasants were able to repay debts or finance expansion from their own resources. And the price of credit was high. For instance, some of the district administration and the government provided seeds on credit that the peasants had to repay with 100% interest (Van Dijk 1940, p. 107). It was generally felt that Hindostanis possessed an economic attitude that enabled them to overcome these barriers.

Table 7.2 Advances granted by Boerenleenbank, by district, 1931–1948

	Nickerie		Saramacca		Suriname		Commewijne	
Year	Number of borrowers	Average sum	Number of borrowers	Average sum	Number of borowers	Average sum	Number of borrowers	Average sum
1931	496	39	76	35	51	4	148	31
1932	487	35	65	39	407	44	64	35
1933	510	35	85	48	138	4	67	62
1934	510	37	76	53	127	1	103	72
1935	526	36	69	44	16	1	139	155
1936	670	38	74	50	22	60	131	638
1937	174	27	74	60	Nb	Nb	115	66
1938	380	24	94	55	5	19	74	65
1939	Unk	Unk	Unk	Unk	Unk	Unk	Unk	Unk
1940	128	19	108	69	9	30	32	128
1941	11	44	73	53	6	18	22	74
1942	7	39	85	44	1	35	20	40
1943	7	54	3	43	62	100	22	111
1944	152	44	Nb	Nb	77	47	21	183
1945	247	46	1	250	72	79	21	175
1946	374	53	72	82	5	80	390	34
1947	816	58	106	69	8	95	1	30
1948	810	58	128	64	70	1063	Nb	Nb

Source: Suriname Report, statistical annexes
Unk unknown

7.4 Unequal Productivity

The start of the rice cycle is highly dependent on Suriname's climate, which consists of four seasons: the major and minor rainy seasons, and the major and minor dry seasons. The rains enabled the cultivation of the heavy clay soil on which rice is planted. These two conditions largely determined the nature of work and the order in which it took place. The problem with rice cultivation was that '…peasants have to wait for the rains to break through properly before planting, which means that cultivation starts at ever-uncertain times for almost all households at the same time; all the work, planting, weeding and harvesting etc., which requires many hands, therefore falls at the same time for all of them, resulting in a repeated severe shortage of labour, which can result in crop losses' (De Vos 1940, p. 133).

The old areas, mainly settlements, had facilities such as dams, canals, sluices and the like. Irrigation was crucial as otherwise in the dry season the clay soil was too hard to cultivate. Conversely, the new areas could suffer from flooding, necessitating a drainage system. In the districts of Nickerie, Coronie and, to a lesser extent, Saramacca, where the polder system came into vogue, water was available in the swamps but its use depended on the construction of dykes, tubes and sluices. In all areas, rainfall determined the ‘farming calendar’, the sequence of activities and the time interval between them, and especially between planting and harvesting (Bürer 1956, pp. 171–174; Hewitt 1922, pp. 14–16; Stahel 1933).

The yield per unit of land (hectare) varied considerably by area as shown in Annex 7.5. The variation includes 2200 kg per ha on Rust and Werk on raw land to double that on Katwijk, while Constantia achieved 2575 kg per ha with ploughing and Frederiksburg (also ploughed) yielded almost double. In 24 of the 40 trials (pieces of land planted with crops meant for measurements), the land was not tilled. In a description of the labour process at the beginning of the 1930s, experts pointed out that the large variation was mainly caused by the presence of irrigation and drainage and the availability of cattle (Snellen 1933, pp. 148–152; Stahel 1933, pp. 11–21). The variation in planting methods was related to the tillage possibilities of the land and the labour required. This correlation persisted until after the second half of the twentieth century.

There may be another factor that explains the variation in yield, namely the physical condition of the farming population. In the 1920s, the diseases and malnutrition of the immigrants were pointed out as reducing their productivity (Lampe 1927; Leys 1920b, pp. 129–121; Van Lier 1977, p. 221ff). So widespread and far-reaching were these diseases that an economist of the Dutch Chamber of Commerce wrote after a visit to Suriname in 1926: ‘The people are sick through and through, and what is already too sick and too weak to work sinks down to Paramaribo’ (De Cock Buning 1926, p. 23). This may sound exaggerated because most peasants continued to live in the districts, but, together with the findings of the aforementioned authors, it shows that there was a serious health problem in the colony. Assuming that physical conditions were unevenly distributed regionally and by ethnic group, this may partly explain the uneven labour input and yield of the rice-growing population.

7.4.1 The Production Process

The rainy season, which fell between mid-April and mid-August, was the period when rice was planted. Some preparations were made for this purpose: repairing the dykes, weeding and tilling the sowing bed. The paddy was first sown and after about 5–6 weeks transplanted into the sawas as shown in Photo 7.3 .

Planting out was only possible when the soil had been softened by the water and turned. The digging was done in very different ways. In the most primitive way, it was done using a tjap (hand hoe), while in the most advanced form, tractors were used. In the 1930s, the use of draft animals increased for both digging and harrowing the land (see Photo 7.4).

Harvesting took place in October. After the paddy was dried and threshed, it was put into sacks, stored and transported to the peel mills. The time needed for threshing and transport could be shortened by the use of draft animals. A labour analysis conducted in the mid-1950s on rice farms where only manual labour was used

Photo 7.3 Hindostani woman in Commewijne planting bibit (young rice plants) at large distances from each other, 1947

showed that a total of 1160 h were needed to grow 1 hectare of rice. The composition of this figure is summarised in the table below.

Table 7.3 shows that harvesting required the largest labour input (310 hours). That period was set at 2 weeks as otherwise the risk of 'suncrack' (breaks in the grain due to too long exposure to the sun) increases. Assuming that the peasant works 6 days a week and 8 h a day, this means a disposition of 96 hours during those 2 weeks. The harvesting capacity, therefore, was 96/310 = 0.31 hectare per artisan. Under these conditions, a household consisting of four adults could harvest a maximum of 1.24 hectares of paddy. This was a considerably smaller area than the 4 hectares advocated by Leys in 1921 (Leys 1921).

When this capacity is compared with the situation in the early twentieth century, when plots were 1 to 1.5 ha in size, technology was considerably more primitive and the number of families was much lower, then the staggering conclusion is that peasants, despite having access to large tracts of land, must have lived far below a subsistence level.

Larger acreage harvested meant peasants had to rely on their neighbours or relatives, provided they did not face the same shortages. But that was usually the case.

Photo 7.4 Ploughing of paddy fields with buffalos, 1947

Table 7.3 Labour in hours required for growing 1 hectare of rice

Weeding and repairing dams and trenches	80
Preparation of seed bed (5 acres) and sowing	34
Weeding of planting field	180
Bibit pulling and planting	248
Suppling	100
General maintenance work	100
Cutting, harvesting/binding	310
Mounding/threshing and winnowing	92
Emballing and storage	16
Total	1160

Source: Bürer 1956, p. 171

De Vos (1940) made it clear that all peasants had to wait for the rains to break through and start the same work at the same time. This situation meant that help from neighbours and others was limited unless offered outside the two-week time interval. The increase in work ability was possible only when there were more hands or when people worked with advanced resources. This was sometimes possible by employing combines for harvesting (which was available only in the Nickerie district), but planting out required more people. The peak labour input in rice cultivation could (then) only be negated by using new implements or employing more people.

Former Minister of Agriculture, Livestock and Fisheries (LVV) Stanley Raghoebarsing pointed out that 'this peak can be flattened if not everyone is sowing at the same time'. As long as the production was not primarily for the market, but for own use, it must have been fine within the margins since the price of paddy did not play any role. In the case of production for the market, all peasants want to be the first to sow so that they are also the first to harvest, because only then do the peel mills pay the best prices (e-mail communication 9 April 2020). However, this development presupposes a high degree of commercialisation, which did not start until after World War II. By then, urbanisation and emigration had significantly depleted the peasant population (see Chap. 10).

Increasing rice acreage made little sense without first removing restrictions on harvesting capacity. This could be done in a couple of ways. First, there was the possibility that larger households could muster larger amounts of labour and thus harvest a larger area. Second, more efficient means of production could be employed, especially labour-saving inputs such as draft animals and combines that increased harvesting capacity. Third, one could use labourers, which presupposed savings or access to credit to pay wages. The latter possibility was rarely mentioned before World War II. And fourth, one could harvest outside the fortnight. This meant poorer-quality paddy, but it was probably the most common solution of the poor rice peasant.

7.4.2 *Land Productivity*

Increased production for the same amount of cultivated land could only be obtained by increasing land productivity. One way to achieve this was to plant closer together, which gave the peasant more paddy per unit of land. These yields differed between districts as well as within a district. Annex 7.5 contains the results of several experiments planting the same rice variety, namely skrivimankoti, reported in 1941. These show that the yield ranged from 2000 to 5025 kg of paddy per hectare. Out of 40 observations, the yield was above 3000 kg in 27 cases and above 4000 kg in 11 cases. These results show no relationship between yield and land cultivation. Whether the land was uncultivated, ploughed or tilled did not matter for the yield.

Differences in yield were also mentioned for an earlier period. Hewitt (1922, p. 16) reported for Nickerie a yield of '1500 kg per field. In good years the yield is considerably higher even'. According to various definitions, a field is 0.05 ha. Converting this figure would then give an unlikely yield of 20 x 1500 kg = 30,000 kg. This is probably a typo and the yield per field should be 150 kg, thus giving the more realistic figure of 3000 kg per ha as the outcome. Ten years later, according to Stahel (1933, p. 17), an unploughed sawa, even one in Nickerie, yielded 2940 kg and a ploughed one 5040 kg of paddy. In the early 1940s, an average of over 3000 kg of paddy per 'deficiently' cultivated hectare was quoted for the same district (Schütz 1941, p. 92). Brandon (1933, p. 89), following an experiment in favour of machine agriculture, reported a yield of 3506 kg in the Commewijne district towards the end of the 1920s. In the 1940s, De Vos (1940, p. 133) reported a yield of 3000 kg per ha. A decade and a half later, Bürer (1956, p. 199) mentioned an average yield of 3000 kg. Yields thus ranged from 3000 to 5000 kg of paddy per ha. In official statistics, however, a general average of 2500 kg per ha was maintained.

The foregoing figures allow only one conclusion, namely that with similar conditions, the natural productivity of the land varied greatly. These conditions include fertility, sunshine hours and rainfall; infrastructure, including water control; rice varieties and seed quality. As it is not clear whether these natural conditions were the same in all experimental fields, certainly before World War II, natural conditions seem to be decisive. It can be assumed that land yields in the Nickerie district were higher than in the other areas. The Stahel and Bürer calculations hover around 3000 kg per hectare. Because these two estimates fall in the period covered by the present analysis, despite the official estimate of 2500 kg, this study uses 3000 kg per hectare.

Exactly what effect irrigation and extension had on production is difficult to determine. Assuming that these factors have improved land productivity, it should be pointed out that labour productivity should also increase. The two are often closely interrelated, for example when denser planting or better water management produces a greater amount of paddy per ha, and thus more labour is needed to harvest it. But this increase in yield does not keep pace with increased labour requirement. If labour productivity stagnates, this leads to a loss of land yield.

7.4.3 Labour Productivity

The availability of labour was initially directly dependent on household size. Around 1939, the agronomist and later Minister of Agriculture, Livestock and Fisheries W.E. Hewitt began a detailed budget survey of 300 peasants. In anticipation of the completion of the survey, some of the findings were published in the 1941 Annual Report of the Department of Agriculture-Economic Affairs (Hewitt 1941). The research was never published in its entirety and the material could not be traced. In 1952, Van Beukering, then director of the Department of Agriculture, Livestock and Fisheries, published material on peasants that was very reminiscent of Hewitt's publication in terms of categorisation.

Table 7.4 has been edited on the basis of data published by Van Beukering (1952, p. 147). However, it is not clear how the sample was composed, which areas were included in the study and what products these peasants grew. Van Beukering (1952, p. 148) does mention that 'a rigid comparison cannot be made from the foregoing

Table 7.4 Size and composition of peasants' households, 1940

Ethnic group	Size of households		Working part		Number of households
	Spread	Average	Spread	Average	
Javanese	3–4.7	4.0	2.0–3.6	2.4	97
British Indians	5–10.0	6.9	2.4–5.0	3.5	105
Creoles	5–6.0	5.4	1.8–3.4	2.7	31

Source: Adapted from Van Beukering (1952, p. 147)

compiled material, because the figures cannot be compared with one another as the analyses were made at various periods when the price levels were not uniform'. This means that the data in Table 7.4 should be understood as a collection of small samples.

Table 7.4 plots household size and working share by ethnic group. The table shows that British Indian households ranged from five to ten members, with an average of 6.9. Javanese and Creole households were smaller. Remarkably, the minimum size among Hindostani households was also as high as among Creoles and higher than among Javanese. On average, half of the British Indian households worked. This was also the case among Creoles. But the working proportion of British Indians was higher on average than that of the other groups. The table indicates not only differences in the quantity of family labour but also differences in labour productivity. Two aspects are particularly important in rice farming. The first is that British Indians used sickles for harvesting, while Javanese adopted this tool at a slow pace. The second aspect was that the British Indians used ploughs and draft animals to a much greater extent (LEZ 1936/1937, p. 150; De Vries 1965, pp. 116, 187). The expansion in both the 1930s and 1940s of the area under cultivation and the harvesting capacity was therefore greater among Hindostanis than other rice growers. Their modernised agricultural capital was more productive.

The decline in the social cohesion of Hindostanis due to modernisation was hardly noticed. It has, however, often been mentioned that Hindostanis are individualistic, competitive, avaricious and the like, but these traits have mainly been mentioned as characteristics of the cultural group rather than as features of a peasantry (Beck 1924; De Klerk 1941; Snellen 1933). Since these traits explain disintegration rather than cohesion, I am inclined to interpret them as characteristics of emerging agrarian entrepreneurs in a modernising agriculture in which households are becoming increasingly independent. No wonder that in the Kielstra period, and thereafter, inter-ethnic cooperation stagnated. Similar stagnation also emerges from Kruijer's (1949) description. His observations, judging from the references, covered roughly the period 1920–1948. 'In those days, people often harvested and planted together or built a house together. Although much of this neighbourly help has disappeared nowadays, it is by no means completely absent. Neighbours still work together in the fields' (Kruijer 1949, p. 215).

Thanks to increased productivity inputs, rice production increased by more than 2.5 times from the 1930s to the late 1940s (Annex 7.1). During World War II, there was some stagnation in production. Exports also collapsed and remained at a low level: the volume exported between the years 1940 and 1948 did not exceed that of previous years. Despite increases in production and productivity factors, rice production and exports continued to disappoint.

It was in the light of these developments that there were repeated calls to interest Dutch peasants in Suriname. There was a belief among agriculturalists that these peasants would do better in the tropics than the Hindostanis. The idea was to let them conduct both agriculture and cattle breeding. This had to be done in the most modern way to optimise both land and labour productivity. Marketing in Europe was also considered benevolent (Dijkhuis 1947; Staal 1919; Verkade-Cartier van Dissel 1937). This thought would be further explored after World War II.

7.5 Processing and Marketing

A major factor causing the rice trade to stagnate was the habit of many peasants of cultivating several varieties of paddy at the same time, which resulted in different varieties of rice being delivered to rice paddy mills. The grains of some varieties were weak. This caused rice grains to break during husking. Technical problems, such as wrong tuning of the machines, were found to promote grain breakage while the multitude of rice varieties gave different colours to the final product. To improve this, the Department of Agriculture started distributing pure seeds to be paid in kind. A compulsory inspection of rice was also introduced in 1932, initially only for export.

The hulling mills differed enormously in terms of capabilities and quality. This industry tended towards overcapacity, so the quality of the rice delivered was partly affected by the fierce competition between the husking mills (Glans 1963, pp. 4–8; LEZ 1931/1932, p. 57). Table 7.5 summarises the development of the number of rice husk mills in the period 1936–1949. It shows that the number of rice paddy mills decreased in Paramaribo and sometimes more than doubled in other districts, with the exception of Coronie. The decline between the years 1942 and 1944 was probably due to the migration of rice peasants to the city.

The control of breakage and the homogenisation of rice varieties were important for foreign trade. In the 1920s, the colonial administration had already observed that surpluses caused large fluctuations in the price of paddy (LPS 1924/1927, p. 15). Domestic production was protected by an import tax of f3.50 per 100 kg that was in force until 1930. But exporting the rice surplus remained a persistent problem because of substandard quality. In the 1930s, the administration sought to gain a foothold in foreign markets. Those markets, so-called 'spot markets', consisted mainly of the Netherlands Antilles and some surrounding areas. In 1946, Suriname joined the International Emergency Food Council, which designated Curaçao and Latin and Central America, with the exception of Cuba and the European colonies, as outlets. As a result, Curaçao was used as a transit country for the export of Suriname's rice (K.v.K 1945/1946, pp. 34–35).

Table 7.5 Number of rice paddy mills by district, 1936–1949

Year	Paramaribo	Suriname	Commewijne	Saramacca	Nickerie	Coronie	Total
1936	–	50	19	9	25	5	108
1937	4	51	18	11	28	5	117
1938	3	56	13	12	31	5	120
1939	3	54	21	19	32	5	140
1940	3	54	26	20	32	5	140
1941	3	54	26	20	32	5	140
1942	3	54	26	20	32	5	140
1943	2	54	25	19	30	5	135
1944	1	49	26	19	31	5	131
1945	1	48	26	21	34	4	134
1946	1	48	29	20	42	4	144
1947	2	55	32	24	45	6	165
1948	1	70	29	26	48	6	181
1949	–	72	33	28	50	5	189

Source: Suriname Report, 1950, p. 28
Note: In the totals of the last three years, one mill that was located in Marowijne is missing and has been omitted here. In all previous years, Marowijne owned no rice mill

World War II diminished the orientation abroad. Anticipating food shortages, the government encouraged the population to buy as much domestically produced food as possible, while peasants were encouraged to expand their production (LEZ 1938/1939, pp. 20–22). Planting material was provided for this purpose and the government promised to buy up any surpluses. Exports declined due to reduced transport possibilities, however. There was some recovery after the war, but rice exports remained stagnant mainly because of the old problems.

The ethnic character of rice production led to various other activities related to processing, transport, storage, packaging, etc. On this, De Vries (1965, p. 120) writes that 'several well-to-do Hindostanis took advantage of the lack of organisation and insight that existed everywhere by acting as moneylenders, buyers and transporters, which gave them the opportunity to act as price pushers for large personal profits. Some Hindostani peasants managed to cope with this, by establishing cooperative rice mills … However, a large proportion of peasants remained dependent on the arbitrariness of middlemen'. Thus, in the rice chain, occupational groups consisting of people of the same ethnic background but with opposing interests emerged. This differentiation, besides modernising the production process, also undermined the internal cohesion of the Hindostani community.

The rice-derived activity was most strongly expressed in rice paddy mills. Again, it was Hindostanis who dominated these related activities. Table 7.6 shows the predominance of Hindostanis in the rice paddy milling sector. The Hindostani rice paddy mill owners numbered 151 out of a total of 189. The main rice districts, Suriname and Nickerie, had 72 and 50 rice paddy mills, respectively.

Transport, especially over long distances, was mainly carried out using donkey carts and korjaals that were also largely owned by Hindostani entrepreneurs (Fokken

Table 7.6 Permit holders of rice paddy mills by ethnicity and district, 1949

Ethnic group	Suriname	Commewijne	Saramacca	Nickerie	Coronie	Total
Creoles	7	1	2	3	3	16
Hindostanis	63	23	23	42	1	151
Chinese	1	2	2	2	1	8
Europeans	1	1	–	1	–	3
Others	–	6	1	2	1	11
Total	72	33	28	50	5	189

Source: Surinaamsch Verslag (Colonial Reports), 1950, p. 27
Note: The Marowijne district had one rice paddy mill in the 'Others' category and is omitted here

2018, p. 235ff; Leys 1925, p. 450). This concentration reflected the fact that rural entrepreneurs, both in rice production and related activities, consisted mainly of members of the Hindostanis.

7.6 Poverty

The most striking feature in the history of the Hindostani peasant was not the abundance of land but poverty, which had several causes. In the previous chapters, I pointed out the unequal starting conditions, such as knowledge, investment, quality, size of plot, location vis-à-vis local markets, health, availability of family labour and the like. These disparities were already emerging on the plantations and translated into inequalities among the starting peasants. I have touched on this poverty in several places in the above. Here I dwell on it a little longer as it was not just a facet of peasant life at the beginning of colonisation.

Peasant poverty has not been a subject of policy at all. This is not to say that individual agronomists did not pay attention to it. One notable exception was Leys, who as early as in 1921 (p. 39) wrote: 'Poverty is the part of the small peasant and will remain so until his death. Husband, wife and children lack the bare necessities all their lives and for all their suffering they are disrespected by the majority of people and no one reaches out a hand to help them, but they do reach out for their produce to get it even cheaper than it is.' Poverty remained a loyal companion of the peasants in the decades that followed, although specific social categories managed to escape it. But the distribution of land ownership, the quality of the land and the limited labour productivity make it clear that the majority of peasants lived in harsh conditions. The only explanation for the absence of mass mortality among peasants is that access to land allowed them to grow their own food.

After World War I, agronomists, notably Leys, Beck, Brandon and De Cock Buning, reported regularly on the hopelessness of peasant life. But the various committees and rapporteurs hardly paid any attention to it. Full attention was paid to business-economic and agronomic aspects of farming: plot size, cultivation method, seed selection, method of planting, tillage, method of harvesting, quality of rice, animal husbandry and the like. This optic was mainly inspired by the need to bring small farming to fruition with which the domestic food supply could be secured and any surpluses exported (Menkman 1940; Stahel 1933; De Vos 1940).

At the beginning of World War II, an agrarian engineer, H.N. van Dijk, who had an eye for peasant poverty, quantified the income and expenditure of Nickerian small peasants. ‘A large part – and this includes the majority of Creole peasants – run their rice culture as a pure pauper culture’ (Van Dijk 1940, p. 107). His calculation of these households omitted some items, however. Below is a full list of the items from his calculation. Van Dijk’s starting point was a plot of 1.4 to 1.5 ha for which a rent of ten guilders per ha was paid with the government providing irrigation and discharge. The farming household should ‘not be too big’. The yield per ha was 2800 kilos of paddy (unhulled rice). All work was done by the peasant at their own expense.

Yield 2800 kg paddy (× 1.5)	4200 = 60 bags of 70 kg
50 kg seed borrowed, 100 kg returned	100
Remainder	4100
Transport to the mill	Dfl. 6
Taxes	Dfl. 15
Rent of the land	Dfl. 10
Total	Dfl. 31

To acquire the amount of f 31 and assuming a price f1.60 per bag, the peasant had to sell 19.4 bags of 70 kg = 1358 kg of paddy. They were therefore left with 4100–1358 kg = 2742 kg of paddy. From that, they could at most have 16 bags of rice made. The cost involved in husking was about two sacks of rice, leaving the peasant with 14 sacks of husked rice to take home ‘… from which they must live (Map 7.1).

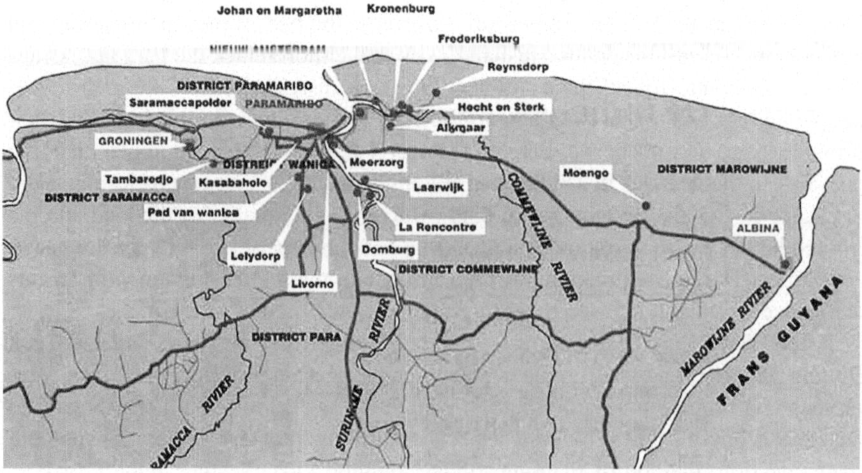

Map 7.1 Drawing showing some settlements. (Source: Ehrenburg and Meyer 2015: 203)

If there is still a father or mother or other relative to support, there is not much left by the next harvest. I know several families who cannot get by on one bag of rice a month' (Van Dijk 1940, p. 108).

To the best of my knowledge, van Dijk was the first agronomist to attempt to quantify the income and expenses of the small peasant. It is notable, however, that he assumed that the entire plot was planted with paddy, leaving no room for the cultivation of vegetables, fruit, flowers and other plants for own use and for keeping cows, sheep, goats, chickens, ducks and the like and for what he called 'side cultures' (Van Dijk 1940, Ch. VI). If these activities are discounted, the area to be cultivated is less than 1.5 ha.

Also noteworthy is the fact that van Dijk does not mention the loss of yield due to insect infestation, which experts estimate to be 15% before the mid-1940s (Abrahamse 1959, p. 55), the distance to the market and the cost of transporting the rice to the consumer, the high interest rates demanded by private lenders (the government was already demanding 100% interest on seed) and the setbacks due to excessive drought and rain. Given these aspects, the conclusion is that Van Dijk's figure relates to a farm larger than 1.5 ha and that, despite this, poverty was a permanent and defining feature of peasant farming. The historical continuity of mass poverty was undeniable.

The issue of poverty was raised again in a lecture on the influence of mechanisation on the development of agriculture in Suriname in 1944 delivered by Captain E. van Dijk (no relation to the engineer Van Dijk). His focus was on large-scale agriculture. On smallholder agriculture, which was characterised by a pauper culture, he remarked: 'By far the largest part of small peasants pay no income tax, no or little school fees, almost nothing for medical treatment and can barely pay the ground rent in the case of a bad harvest' (quoted in Van de Walle (1945, p. 72)). Van de Walle contested the prevailing view that the rice culture produced little more than a pauper's existence, as authors such as Van Dijk (1940, p. 107) and De Klerk (1941, p. 104) had argued not long before. He did so by pointing to flourishing Hindostani entrepreneurs during the war years, in particular the more than one million in real estate transactions in 1943 (Van de Walle 1945, p. 73). However, his argument is not convincing because he did not adequately identify that these were two different segments of the Hindostani community: the small peasants who lived in the districts and the entrepreneurs who mostly lived in the city or its surroundings.

De Klerk (1941, p. 104) argued that rice exports had created a hunger for land among the Hindostanis, '... and it is a gratifying fact that the government, in particular Governor Kielstra and the district commissioners De Rooy and Van Assenderp, have tried to satisfy this hunger everywhere by creating new polders'. This development applied to all districts, especially to the Nickerie district, and the polders of Paradise, Longmay, Nieuw Waldeck and Sidoredjo. The Second World War undermined the cultural stability in these regions as more and more non-agricultural employment emerged. In a publication in the mid-1940s, Kielstra (1946, p. 355) lamented: 'While on the one hand, the presence of American troops facilitated the sale of agricultural produce, on the other, the agricultural centres became disorganised by the disproportionately high wages paid by the Americans.'

Van Beukering presented data that also led to the conclusion that peasants were leaving their plots. In 1952, he published an article based on unpublished research conducted from 1941 to 1946 by the Department of Agriculture. The sample consisted of Javanese, Hindostani and Creole households. Below is a compressed view of the table showing only the sources of income of the 105 Hindostani households. Non-agricultural earnings, as Van Beukering (1952, p. 146) outlines, were for supplementing household income.

Of the seven sites, the bottom four are located in the Nickerie district. With the exception of the Corantijnpolder, the plots in the polders are larger. Strikingly, the Nickerie polders also had a larger percentage of the available area under cultivation, with all being above 80%. Total incomes were also higher, but the proportion sourced from outside agriculture varied. In the Hamptoncourt and Boonackerpolder, the percentages were 6 and 8, respectively (Table 7.7).

but in the other polders, the percentage of externally earned income hovered around a quarter of the annual income. At the settlements of Johan and Margaretha and Bruinendaal (in the Commewijne district), where the available land was considerably smaller, more than half of the income came from outside agriculture. Based on this table, three conclusions can be drawn. The first is that plots of about 4 ha, as was assumed to be reasonable in policy circles, were insufficient for a family to make ends meet. All families required supplementary income, ranging from 8 to almost 60%, from outside agriculture. It is therefore surprising that Van Beukering (1952, p. 148) concluded on the basis of these figures: 'The self-supporting small farm under Surinamese circumstances and with the outlined agricultural system would apparently be around 4 ha.'

Table 7.7 Area and incomes of British Indian households during World War II; in brackets the percentage of land under cultivation

Place	Available land in ha	Ha cultivated	Income from agriculture (1)	Income from outside agriculture (2)	Total income (3)	2/3 als %
Johan en Margaretha	1.6	0.93 (0.58)	81	115	196	59
Bruinendaal	1.64	1.12 (0.68)	134	138	272	51
Saramaccapolder	2.91	2.12 (0.73)	156	53	209	25
Van Drimmelenpolder	4.26	3.55 (0.83)	353	113	466	24
Corantijnpolder	2.00	1.61 (0.81)	147	53	200	27
Hamptoncourtpolder	5.24	4.45 (0.85)	960	56	1016	6
Boonackerpolder	6.60	6.11 (0.93)	1035	96	1131	8

Source: Van Beukering 1952, p. 147

Much later, Van Dusseldorp (1967, p, 127), basing himself on an evaluation of a small place named La Poule located in the district of Saramacca, argued that a farm of 4 ha, of which 2 ha was planted with rice and cultivated by a family that could yield 45 man-days per month, would not need external labourers. That was at odds with previous data, according to which the 4 ha-sized peasant farm was not big enough. Not to mention that the labour of the peasant household was insufficient to work the land. Most likely Van Dusseldorp assumed a higher labour productivity than his predecessors.

The second conclusion is that there was a high degree of income inequality within the Hindostani farming community. Peasants in the Boonackerpolder earned almost six times the income of those in Johan and Margaretha. This disparity probably varied by district. Arguably, within the districts, differences in productivity and location also contributed to the large disparities in income. This conclusion is based on production-side factors, but other conditions, especially the volume of products that could be sold, transport costs, the role of middlemen and the prices obtained, are also important determinants in establishing the level of household income. As mentioned earlier, those sales were highly dependent upon proximity to markets and infrastructure. Land ownership alone is thus an insufficient indicator to give an idea of the peasants' wealth, as has often been done in the recent literature (Choenni 2016, p. 469ff; Fokken 2018, p. 216ff).

Thirdly, it can be inferred from the table that farming was a part-time existence for certain segments. The average size of the plots varied by location, but so did the percentage under cultivation. There is no clear relationship between plot size and the percentage that was cultivated. The only clear relationship in the table is that the smaller the available land, the larger the share of income earned outside agriculture. In particular, households in Johan and Margaret and Bruinendaal earned more than half of their income outside agriculture. This trend had already been reported in the first decades of the twentieth century (McNeill and Lal 1915). Agriculture had become less and less interesting for peasants and their main source of income gradually shifted to off-farm activities.

Persistent poverty led to the departure of peasants and their children to other places. In some districts, the number of peasants increased. This can be explained by natural population growth, marriages and related relocations, and the influx of new migrants, especially in the border districts of Nickerie and Marowijne. But the area around the city of Paramaribo actually saw a decline in the agricultural population (Van Dusseldorp 1963). In contrast, the number of occupations derived from agriculture increased. To make this urbanisation possible, the peasant had to master Sranan (the lingua franca), have a significant surplus, be physically strong and have the means to transport and sell their wares over long distances (De Klerk 1941; Kruijer 1951). As a rule, such geographical and social mobility was reserved for the physically strong. Thus, those left behind in agriculture were mainly the old and weak, while the young increasingly, despite living in the districts, sought their sources of income outside agriculture.

7.7 Conclusions

Kielstra's taking of office marked a different type of agricultural policy. It was less aimed at restoring or compensating plantation agriculture or at merely increasing production, but rather mainly at strengthening social cohesion among small Girmitiya peasants so that together they would bear the infrastructural burden and achieve a higher level of production. He legally founded this philosophy by setting up village municipalities that had to take care of communal affairs, however. Collective interest representation was not an unqualified success: in Nickerie, cooperative associations did come into being, especially in the infrastructural sphere such as the construction of polders, sluices and the laying of tubes, but in the other districts, such cooperation was virtually absent.

What did persist was the traditional cooperation between neighbours and relatives during periods of planting and harvesting. However, the mutual support became less and less necessary, firstly because family formation was booming, which allowed the Girmitiya peasants to deploy their own family labour, and secondly because the peasants were carrying out the same activities at the same time, which limited their ability to cooperate. Thus, the compulsion to plant and harvest within a limited time, on the one hand, and the modernisation of rice farming, on the other, affected traditional ethnic cooperation negatively. Cooperation was easy when peasants owned small plots. What this conclusion underscores is that, because of the relatively small scale on which they operated, peasants finished their own farming activities within a short period of time to be available for group members.

Kielstra's ruralisation policy was not aimed at this traditional ethnic cooperation that was so typical for moral communities but at a form of cooperative that could address major infrastructural needs. If these were addressed, it was short-lived. In Nickerie in particular, attempts were made to encourage cooperatives by issuing favourable titles to land, but these proved insufficient as an incentive. At the same time, migration to the city continued. Governor Kielstra's ruralisation policy aimed at keeping peasants in agriculture was not very effective.

This was not so much due to the governor's policy as to various other factors. The most striking was that, despite statements by agronomists, successive administration regimes paid scant attention to the peasants' long-standing debilitating poverty. This is not just a moral observation. Peasants' poverty undermined their labour commitment and limited the use of resources (such as cattle) to increase agricultural productivity. Kielstra fell in the midst of a development in which this paucity was already far advanced, but he paid little attention to it. His policy efforts focused on improving the social and physical infrastructure by establishing village communities and looking after collective interests—not on fighting poverty. In doing so, he put all his cards on the stronger peasants in the Nickerie district. That was where the increase in production was taking place and where investments were promising. However, he underestimated the undermining effects of poverty (Map 7.2).

However promising rice production in Nickerie was, it was not yet such that its export would be macroeconomically significant. It was therefore no coincidence

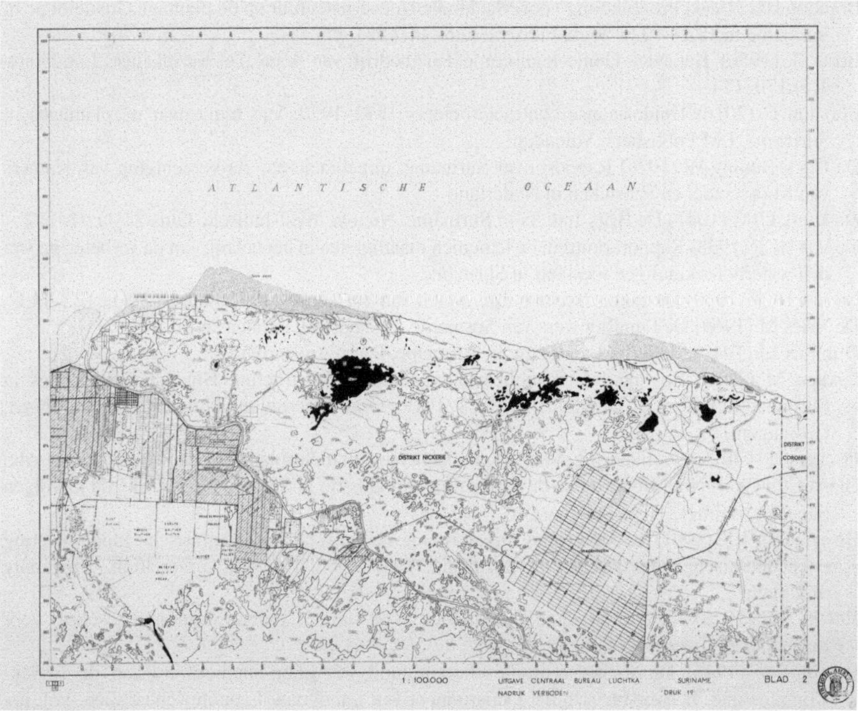

Map 7.2 Survey soil map of the coastal strip of the Nickerie district (1977). (Source: Morenc 1988, p. xiii)

that old ideas of colonising Dutch peasants in Suriname (and allocating them much larger tracts of land) were revived. At the end of World War II, Suriname was preparing for a large-scale intervention to revive agriculture, but this was mainly due to international developments.

References

Aas S (1980) The relevance of Chayanov's Macro Theory to the case of Java. In: Hobsbawn EJ, Kula W, Mitra A et al (eds) Peasants in history: Essays in honour of Daniel Thorner, eds. Oxford University Press, Calcutta, pp 221–248

Abrahamse J (1959) Padi: Verslag van de in 1958 Gehouden Opnamen Betreffende de Oppervlakte en de Produktie. Ministerie van LVV, Paramaribo

Adhin JH (1969) Ontstaan en Ontwikkeling van de zg. Aziatische Huwelijkswetgeving; overdruk uit: Een Eeuw Surinaamse Codificatie. Paramaribo

Beck S (1924) De Creolen en de Klein–Landbouw. In: Beck S (ed) Lezingen over Surinaamsche Problemen. Kersten, Paramaribo, pp 1–34

Boeke JH (1930) Dualistische Economie. Van Doesburgh, Oratie Leiden

Bosz AJAQ, Albrecht AJ (1954) Drie Eeuwen Grondpolitiek in Suriname. Van Gorcum & Comp, Assen

Brandon HG (1933) Proefneming voor Machinale Hoeverijstculuur op de plantage Guadeloupe in Suriname in 1925–1926. West-Indische Gids 14(1):81–98
Bürer T (1956) Het Niet–Gemechaniseerde Rijstbedrijf van 4 ha. De Surinaamse Landbouw 4(5):170–174
Choenni C (2016) Hindostaanse contractarbeiders 1873–1920. Van India naar de plantages in Suriname. LM Publishers, Volendam
De Cock Buning W (1926) Rapport over Suriname, uitgebracht aan de Vereeniging van Kamers van Koophandel en Fabrieken in Nederland
De Klerk CJM (1941) De Brits-Indiërs in Suriname. Nieuwe West-Indische Gids 24(1):97–117
De Vos HCP (1938) Rapport omtrent de te nemen maatregelen in het belang van de verbetering van den waterbouwkundigen toestand in Suriname
De Vos HCP (1940) De economische hulpbronnen van Suriname. De Economist 89(1):122–143
De Vries M (1965) De Landbouwers van Suriname. Planbureau, Paramaribo
Dijkhuis LJ (1947) Moderne Landbouw in Suriname. Paris, Amsterdam
Fokken M (2018) Beyond being koelies and kantráki. Constructing Hindostani identities in Suriname in the era of indenture 1873–1921. Doctoral dissertation, University of Groningen, Groningen
Frickers J (1939) Vleeschkeuring, Vleesch en Slachtveeteelt in Suriname. Eben Haezer, Paramaribo
Glans CL (1963) Nota Betreffende de Rijstpelmolenindustrie in Suriname en Inzake de te Volgen Saneringspolitiek. LVV, Paramaribo
Gowricharn RS (1991) Economische transformatie en de staat. Over agrarische modernisering en economische ontwikkeling in Suriname, 1930–1960. Doctoral dissertation, University of Utrecht
Hewitt WE (1922) Nickerie als landbouwdistrict. West-Indië, Landbouwkundig Tijdschrift voor Suriname en Curaçao 7(4):5–20
Hewitt WE (1941) De sociaal-economische toestand van klein-landbouwers aan de Boven-Commewijne. Jaarverslag van het Departement van Landbouw-Economische Zaken, Vol. not reported, p 73–87
Junker L (1940) Het irrigatie vraagstuk in Suriname. West-Indische Gids dl 23(22):226–233
Kielstra JC (1918) Recht en Werkelijkheid. Inaugurele rede, Landbouw Hogeschool Wageningen
Kielstra JC (1927) Nieuwe mogelijkheden voor Suriname. De Economist 76(1):181–196
Kielstra JC (1946) Application of Netherlands Indies agrarian principles to Surinam (Netherlands Guyana). In: Caribbean land tenure symposium, Washington, pp 335–347
Kruijer G (1949) Landbouw-Cooperatie onder de Aziatische bevolkingsgroepen van Suriname en Brits-Guyana. Nieuwe West-Indische Gids 30(1):209–235
Kruijer G (1951) Urbanisme in Suriname. Tijdschrift van het Koninklijk Nederlands Aardrijkskundig Genootschap, jrg 58:31–63
Lampe PHJ (1927) Suriname. Sociaal–Hygiënische Beschouwingen. De Bussy, Amsterdam
Leys JJ (1920a) Het Arbeidersvraagstuk. West-Indië 3:114–121
Leys JJ (1920b) Onze Veestapel. West-Indië, landbouwkundig tijdschrift voor Suriname en Curaçao 2(2):69–75
Leys JJ (1921) De Kleine Landbouwers en het Klein Landbouwbedrijf. West-Indië, Landbouwkundig Tijdschrift voor Suriname en Curaçao 6(1):31–44
Leys JJ (1925) De veeteelt in Suriname. Nieuwe West-Indische Gids 6(1):441–460
LEZ (1941) Jaarverslag van het Department van Landbouw–Economische Zaken. Several years
Liems JA (1918) Is de Creool voorbestemd om dienstbaar te worden aan den Brits–Indiërs? West-Indië, Landbouwkundig Tijdschrift voor Suriname en Curaçao 3(3):140–144
LPS (1924/1927) Jaarverslagen van het Landbouwproefstation Suriname. Agricultural Experimental Station, Published irregular
McNeill J, Lal C (1915) Report to the government of India on the conditions of Indian immigrants in four British colonies and Surinam, Part II. London
Memorie (1919) Memorie van het Comité tot Bevordering van den Middenstandslandbouw in Suriname, op 29 augustus 1919 den Gouverneur Aangeboden. West-Indische Gids 1(2):163–176

Menkman WR (1940) Landbouw-economische politiek in Suriname. West-Indische Gids 23(22):12–19
Morenc J (1988) Surinaamse Kleine Landbouw en Landbouwbeleid: Een Structurele Analyse. Doctoral dissertation, Radboud University of Nijmegen
Ramsoedh H (1990) Suriname 1933–1944: Koloniale politiek en beleid onder gouverneur Kielstra. Doctoral dissertation, University of Utrecht
Schütz HLH (1941) De Ontwikkeling van den Landbouw in Nickerie gedurende de laatste 10 jaar, 88–100. Jaarverslag van het Departement van Landbouw-Economische Zaken, Paramaribo
Snellen E (1933) De Aanvoer van arbeiders voor den landbouw in Suriname. Doctoral dissertation, University of Wageningen
Staal J (1919) Memorie van het Comité tot bevordering van den middenstandslandbouw in Suriname, op 29 augustus 1919 den Gouverneur aangeboden. Overgenomen uit De Surinamer van 2 en 6 november 1919
Stahel G (1933) De Rijstcultuur in Suriname; Amsterdam, z.j. (overdruk uit De Indische Mercuur, 22 en 29 november, en 6 dec 1933), publication of the Landbouwproefstation te Paramaribo, no. 2
Van Beukering JA (1952) Types of farming. Caribbean Economic Review IV(1&2):140–163
Van de Walle J (1945) Suriname, Rapport uitgebracht door het hoofd van den Gouvernements Persdienst in Suriname. Paramaribo
Van Dijk HN (1940) Mechanische Rijstcultuur: Economische en Technische Resultaten Na Zes Jaar Pioniersarbeid te Nieuw–Nickerie, Suriname. Amsterdam: Paris
Van Dusseldorp DBWM (1963) Geografische Mobiliteit en de ontwikkeling van Suriname. Bijdragen tot de Taal-, Land-, en Volkenkunde 199(1):18–55
Van Dusseldorp DBWM (1967) Meerdimensionale Overheidsplanning. De overheidsplanning van Suriname in de periode 1952–1964, getoetst aan een relatiemodel. Doctoral dissertation, Wageningen
Van Lier R (1977) Samenleving in een grensgebied: Een sociaal–historische studie van Suriname. Emmering, Amsterdam
Van Traa A (1946) Suriname 1900–1940. Van Hoeve, Deventer
Van Wengen GD (1972) De Javanen in de Surinaamse Samenleving. Sticusa, Amsterdam
Verkade-Cartier van Dissel EF (1937) De mogelijkheid van landbouwkolonisatie van blanken in Suriname. Doctoral dissertation, University of Utrecht

Part III
From Peasant to Urban Dweller

Chapter 8
Mechanisation of Rice Farming, 1950–1960

Abstract Mechanisation, the replacement of labour by machines, occurred on farms using tractors and combines and was predominantly implemented in the period 1950–1960. Its success or failure can be gauged from the persistence of the smallholder farmer. Using this indicator, the chapter is structured as follows. Section 8.2 discusses the urbanisation that undermined the consolidation of the district population and the gradual transformation of the peasantry. The government responded to this with a new form of ruralisation policy that consisted mainly of the development of district centres. That analysis is followed in Sect. 8.3 by an account of the attempt to revitalise the country by modernising agriculture to make it attractive for farmers. In Sect. 8.4, the planned build-up is followed by the establishment of family-based middle-class settlements in Nickerie. Section 8.5 discusses the effects of mechanisation. That included the erosion of the relationship between the traditional agrarian human capital and the cooperation among the Girmitiyas. Section 8.6 discusses the effects of these interventions on rice production.

Keywords Mechanisation · Urbanisation · Agrarian reforms · Family farms · Ethnic ties

8.1 Introduction

Within agricultural economics, there is a theory that focuses on the 'persistence of the small farmer'. According to this theory, the economic superiority of capitalism is not limited to industry but extends to agriculture as well. In other words, the tendencies towards innovation and concentration of productive resources, including land, also occur in the agricultural sector. The theme of the persistent small farmer was frantically debated by Indian economists after World War II as part of the Green Revolution. The expectation was that the development of agrarian capitalism would produce an agrarian bourgeoisie and a declining class of proletarians (Djurfeldt 1984, pp. 139–140). While this polar development did occur in industry, it appears

R. Gowricharn, *The Girmitiya Peasants in Suriname*,
https://doi.org/10.1007/978-3-031-67961-2_8

that the 'small farmer' has not been outcompeted in capitalist agriculture but persists stubbornly.

Various explanations have been put forward for the survival of the 'small farmer'. These farms were found to have higher returns on average. The small farms were small only in terms of land ownership, but based on other economic indicators such as yield per unit of land, capital or labour, they were found to be among the most productive enterprises. Thus, they had to be highly mechanised farms with high land productivity. This led to a negative statistical relationship that became known as 'the inverse relationship'.

A later study of the transition by the Green Revolution brought together the results of surveys conducted at different time points and in different subregions of India (Patnaik 1976; Roy 1981). These showed that as commercialisation progressed within an area, certain farms were able to enjoy higher yields due to the availability of irrigation and 'higher yield varieties'. The inverse relationship emerged only in cases of uneven land productivity, mostly due to differences in irrigation. In areas where new technologies were widely adopted, the inverse relationship did not occur (Roy 1981, pp. 217–224). Nevertheless, later authors followed the classic interpretation that the inverse relationship would marked the transition from primitive to modern agriculture (Dorward 1999; Dyer 1991). But it is questionable whether this relationship occurred everywhere and whether it mattered. For, despite the persistence of the (modern) smallholder farmer, income could be unsatisfactory because price developments were disregarded.

Even later, two Indian economists re-examined the link between small farms and higher yields (Gaurav and Mishra 2015). Using a large data set and holding various confounding variables constant (such as household, occupation, caste, etc.), they found a negative relationship between farm size and yield. But they did point out that it '…ignores the low absolute levels of their returns, which raise questions about the sustainability of their livelihoods. This is further aggravated by the fact that they (the farmers, RG) pay relatively higher unit costs and because of their greater dependence on purchased inputs' (Gaurav and Mishra 2015, p. 178). So here the issue is not the statistical relationship but the relationship between net income and firm size. To this end, the prices and product mix of firms should be included in the calculation (Dorward 1999).

In Suriname, after World War II, the small farm was equated with a rice-growing and mechanised family farm, which was set at 4 ha. Earlier attempts in the US and British Guiana to mechanise family farms were favourable and had raised the desire to mechanise in Suriname too. Initially, these were relatively small farms operated using tractors and combines, but they required some modifications to make exploitation possible. Because mechanisation was prepared and partly implemented in the period 1950–1960, it is important to verify whether the inverse relationship, which has been observed in several countries, also occurred in Suriname. This would indicate the persistence of the small farmer, i.e. viable small mechanised (family) farms. This period, in which mechanised rice farming boomed, was also a turning point in Suriname's agricultural history. Its success or failure can thus be gauged from the persistence of the smallholder farmer.

The next section discusses the persistent urbanisation that undermined the consolidation of the district population. The government responded to this with a new form of ruralisation policy that consisted mainly of the development of district centres. That analysis is followed in Sect. 8.3 by an account of the attempt to build up the country agriculturally by modernising agriculture to make it attractive for farmers. In Sect. 8.4, the planned build-up is followed by the establishment of family-based middle-class settlements in Nickerie. Sect. 8.5 discusses mechanisation. This development eroded the relationship between the traditional agrarian human capital and the cooperation among the Girmitiyas. Section 8.6 discusses the effects of these interventions on rice production, and the concluding section interprets the findings.

8.2 Urbanisation

Migration to the city is an old phenomenon. In Suriname, even during slavery, many freed slaves settled in and around Paramaribo and this development has never been reversed. Although city dwellers sometimes moved to the districts, Paramaribo remained the attraction. It was no different for Hindostanis. Most of them settled in and around the city because they had professions other than farming: they were traders, transporters, shopkeepers, house owners, sellers of vegetables, milk, eggs, charcoal, meat and fish, and sometimes they worked in white-collar professions, including as teachers and clerks. This migration to the city meant that peasants' children, especially sons, had a declining interest in taking over their parents' farming business. It led to concerns about farm succession and continuity early in the twentieth century (Beck 1924).

In the 1920s and 1930s, the value of bauxite outstripped that of agriculture and other products, at least as measured by export value. On the eve of World War II, American troops were stationed in Suriname to protect bauxite, which was of great importance for warfare. The Dutch army, the Troepenmacht in Suriname (TRIS), was also expanded. Roads were constructed and improved, bridges were built and numerous other infrastructural activities were undertaken (Gowricharn 1991). The opening up of the districts, the jobs created and the increased purchasing power together strengthened the city's pull on the peasantry (and its children). The share of Hindostanis in the city increased, as shown in the table below.

Table 8.1 shows that the relative share of the urban population increased in the period 1938–1947, only to decrease slightly between 1948 and 1952 (last row). The largest increase occurred among the Hindostani population group, followed by the Javanese. The share of Creoles decreased over the years. Nevertheless, they remained by far the largest segment in the urban population.

The distribution of population between districts and city, or between agriculture and other economic sectors, is not as unambiguous as the statistics suggest. In many cases, there appeared to be 'transitional forms' from peasant to (urban) wage labourer. This phenomenon was approached in later studies from different angles, such as part-time farming, the informal sector and marginalisation (Luning 1967,

Table 8.1 Ethnic composition of Paramaribo's population, excluding Maroons and native Indians, 1938–1952, in percent

Census group/year	1938	1947	1952
Europeans	3.0	1.9	2.5
Creoles	79.4	74.7	71.7
Hindostanis	9.1	14.9	16.9
Javanese	1.6	4.1	4.5
Chinese	2.9	2.4	2.7
Others	4.0	2.0	1.8
Paramaribo abs. (1)	52,760	75,233	86,401
Suriname abs. (2)	152,589	181,984	212,230
(1) as % of (2)	34.6	41.3	40.7

Source: Eswin (1956, pp. 7–8)

pp. 99–107; Morenc 1988, p. 124ff; Van Gelder 1984, p. 74ff). The transitional forms were characterised by combining various activities and functions in the same person. For example, a person could be a farmer while being a land labourer or fulfilling another function as his main occupation. Based on the Second Agricultural Census Suriname (TLS 1959), a breakdown can be made of farmers' main and secondary occupations. Table 8.2 represents a condensed view of the transition from district to urban in the late 1950s.

Table 8.2 shows that 68% of farm managers whose main occupation was agriculture had farms smaller than 4 ha. Of those for whom agriculture was a secondary occupation, the figure was 87%. Those with smaller farms were much more likely to practise agriculture as a secondary occupation and thus have a main occupation outside agriculture. Although these people were also employed in agriculture, they had become part-time farmers.

Van Dusseldorp (1967) pointed out that many small farmers continued farming because they were focused on self-sufficiency. 'This form of agriculture, focusing on rice farming, requires only a few months of intensive labour. After that, money can be earned in trade, or in temporary wage employment, to satisfy certain needs, such as the need for a bicycle, moped, sewing machine and the like, or to meet social obligations. That the work pattern described above is very attractive to many agrarians is shown, among other things, by the fact that the secondary profession in Suriname in the agricultural sector takes up a lot of attention and time of the farmers and their families' (Van Dusseldorp 1967, p. 169). However, such a pattern of work presupposes the presence of employment opportunities in the immediate vicinity.

This development varied from district to district. The Second Census of Agriculture Suriname (TLS 1959, p. 43) shows that the number of farm managers whose main occupation was farming and working on plots larger than 0.48 ha in Nickerie was 2036, while in the district of Suriname it was 4751. In the districts of Saramacca and Commewijne, those numbers were 1597 and 1364, respectively. The number of farm managers with farming as a secondary occupation in these four districts was 340, 2376, 229 and 1313, respectively. The trend of part-time farming

Table 8.2 Numbers of farm managers by main and secondary occupation and by size class of plots, 1959 (column percentages in brackets)

Size class parcels	Agrarian business owners	
	Main profession	Side occupation
< 0.48	359	728
0.48–1	912	1107
1–2	2799	1556
2–3	1826	688
3–4	1381	427
< 0.48–4	7277 (68)	4506 (87)
4–5	889	221
5–6	511	94
6–7	437	89
7–8	253	56
8–10	388	66
10–20	618	103
20–50	209	43
4–50	3305 (31)	672 (13)
50–100	39	9
100–200	16	6
200–500	21	7
> 500	24	3
50–> 500	100 (1)	25 (–)
Total	10,682 (100)	5203 (100)

Source: TLS (1959, p. 42)

and hence semi-urbanisation was strongest in the districts of Suriname (2376) and Commewijne (1313). Urbanisation thus did not imply an absolute break with farming but represented a gradual disconnection with that sector. According to the TLS authors, this was linked to the proximity to the city of Paramaribo where alternative employment was highest.

Urbanisation was caused by pull factors in the city. The long-standing small size of plots at settlement sites indicated both constraints on land ownership and hidden unemployment. However, given the large percentages of uncultivated land outside settlement sites, the foremost problem was a lack of labour capacity to cultivate the available land (see Table 6.1). An increasing number of peasant children, the city or its proximity offered the opportunity to make a living as a horticulturist. A study by Kruijer (1951) found that besides economic distress, especially among Hindostanis, other circumstances also contributed to the migration to the city, such as limited job opportunities in the countryside and the desire to continue studying.

These findings were confirmed in a later study (Van Dusseldorp 1963a). Based on the Second General Census held in 1950 and focusing on the difference between place of birth (taking into account immigration) and place of residence, Van Dusseldorp (1963a, pp. 40–41) concluded: 'The urbanisation of Hindostanis began well before the Second World War and reached its peak in the period 1940–1945 …

The impression is that among the Indonesian and Hindostani population the urbanisation process has taken place and is still taking place in stages … In many cases, people do not move directly to the city, but first settle in the immediate vicinity of it and then move into the city at a later date.'

Van Dusseldorp's Appendix 3 (pp. 54–55) shows that the percentage of Hindostanis in the population of Paramaribo increased from 9.7 to 14.4 in the period 1940–1945 and reached 19.9 in 1959. The main causes of the urbanisation, according to Van Dusseldorp, were the large place occupied by the government in the service sector, the concentration of service provision (including unemployment provision) in Paramaribo, the low level of employment in agriculture, the increasing upward social mobility of the district residents and the accompanying occupational differentiation (Van Dusseldorp 1963b, pp. 42–44).

The demographic shift of Hindostanis from district to city was induced not only by the attraction of the city but also by the constraints of life in the districts. The passion for land was not enough to keep Hindostani farmers in the districts, even though the policy goal since Governor Kielstra had taken office until the 1960s had been to ruralise the population (Van Dusseldorp 1963a). The inverse relationship, aimed at the consolidation of the family farm, could not occur because in the surrounding districts, farmers left all or part of their plots. This meant that the persistence of the small farmer, however important in other agricultural communities, lost its relevance. Moreover, the inverse relationship could easily be interpreted as the outcome of intensive cultivation of land by the small farmer, as has been done by some economists (Akram-Lodhi 2001; Ansoms, Verdoodt and Van Ranst 2008). Only under the condition of scarcity of land is the farmer forced to exploit the land intensely. The question is whether this production can be sustained for long without fertiliser.

8.3 Agrarian Reforms

Before World War II, efforts focused on diversification of the traditional export crops (cocoa, coffee, sugar), but after World War II, citrus and bananas and, of course, rice enjoyed increasing interest. In 1947, the Prosperity Fund was set up, which was meant to inventory the natural and human resources and improve production conditions to revive the economy (Adhin 1961). Rice, however, remained the main dish of (agrarian) Suriname. Besides rice cultivation, efforts were also focused on developing animal husbandry and fisheries.

A condition for the development of the agricultural sector was that urbanisation was curbed. It was a continuation of the ruralisation policy of Governor Kielstra, who had set up water boards and village communities for this purpose to give farmers more responsibility, hoping that this would create social communities in the districts. That plan failed. According to Van Dusseldorp (1963a, pp. 19–20), this was due to four factors: the tension between the governor and the local elite; the financially unfavourable conditions in which the settlements found themselves when they were declared village municipalities and were no longer subsidised by the government; the opposition of the district commissioners, who saw their power curtailed by additional demands Governor Kielstra placed on that office; and the inadequate preparation of the population for their new responsibility. In the new scheme of ruralisation, 48 residential centres were developed for all districts, called 'Landelijke Opbouw' (national development), with the necessary (and otherwise minimal) facilities such as a health post, a school and a police station to have an acceptable standard of living in the district (Van Dusseldorp 1963a, pp. 74–75).

The reorganisation of the credit system was part of an overall reshaping of the banking system and was based on a report to that effect (Banking 1948). The Boerenleenbanken were replaced by the Volkscredietbank (People's Credit Bank), to which the 'small farmer' could turn. A special feature of the new system was that loans were now obtained when the appropriate commercial criteria were met. Admittedly, the funds were fed by the government, but the opportunity for the 'small farmer' to obtain credit was limited. The agricultural sector of 'modern Suriname' offered a new vista. Storage and transport facilities were improved, the central market in Paramaribo was expanded and the marketing of agricultural products as well as information to farmers received more government attention. All these measures were aimed at revaluing the agricultural sector and were, in a way, the expression of the post-war development ideology (Gowricharn 1991).

To strengthen the agricultural sector, the agricultural area was expanded and the existing area, fragmented into plots, was improved. Three polders (the Clara Polder, the Nanipolder and the Groot Henar Polder) were created in the fertile part of the coastal plain in the Nickerie district covering an area of over 10,000 hectares. There were also experiments with farms on poorer land such as the Slootwijk plantation in the Commewijne district and at Lelydorp, a settlement near Paramaribo. These experiments were reported separately (Welvaartsfonds 1955, pp. 74–116). Farm size proved to be a persistent problem. Data on farm size are scarce, but 1953 saw the first General Agricultural Census, which was published in part in Courbois (1958). Table 8.3 is based on those results.

Table 8.3 shows that over half of the farms in 1953 were smaller than 2 ha and 78% were even smaller than 4 ha. Together, they accounted for 19% of the available area. The smallest 9400 farms covered 9000 ha, averaging 0.96 ha per farm. In

Table 8.3 The number of holdings and area by size class in hectares, 1953

Size class	Number of holdings	Holdings as %	Total number of ha	As %	Average size
<2	9400	53	9000	8	0.96
2.0–4.0	4400	25	12,000	11	2.73
4.0–8.0	2300	13	12,000	11	5.22
8.0–20.0	1100	6	12,000	11	10.91
>20	450	3	67,000	60	148.89
Total	17,650	100	112,000	100	6.35

Source: Courbois (1958, Appendix 1, 2)

contrast, 60% of the area was occupied by 3% of the number of agricultural enterprises (last row).

Little is known about these enterprises, but most likely they mainly consisted of plantations and other large agricultural enterprises. The small size of the small farms was a barrier to the use of mechanical means of production, especially tractors. Photo 8.1 gives an impression of how vast the rice fields had to be to profitably exploit tractor use. Thus, the tractor was beyond the capability of smallholder farmers.

Six years later the acreage distribution in Table 8.3 appeared to have improved only slightly (TLS 1959). The smallest farms had decreased a little and farms of between 2 and 4 ha had increased slightly in absolute and percentage terms. But the average farm size remained almost the same. In the other size classes too, both in absolute and percentage terms, the ratios did not change much. In other words, the changes in the 1953–1959 timebox were minimal in terms of access to land and can also be explained, for example, by better registration. That is to say, the reform of the agricultural sector had little effect.

The TLS (1959, pp. 56, 58) also showed that the 15,910 agricultural holdings together had cultivated 35,561 ha. A comparison with Table 8.3 shows that the

Photo 8.1 Ploughing a rice field in the Nickeriepolder with a tractor, 1947

number of farms decreased from 17,600 to 15,910 and the number of ha from 112,000 to 37,886 during the period 1953–1959. In 1959, 87% of agricultural holdings consisted of rice farms, which together had covered 28,543 ha or 80% of agricultural land under cultivation. The predominance of rice is also evident from the fact that of the 388 agricultural unions larger than 20 hectares, 294 were rice farms.

Besides the skewed distribution of land and the one-sided concentration on rice, the fragmentation of farms into multiple plots also posed a problem. The TLS shows that land fragmentation was least in the smaller farms. However, the smaller farms not only consisted of fewer parcels but the size of these plots was also smaller on average. This pattern of 'land ownership' was complicated, among other things, by the fact that the legal title to the land (mainly leasehold) was fixed for 75 years. Even if the land was not in use, the legal title remained in force. This made amalgamation of pieces of land impossible. Increasing the productive area was therefore sought in the expansion of land in the new agricultural areas. These were found in the Nickerie district.

8.4 Family Farms

From the colonisation of the British Indian immigrants, the colonial government employed the concept of family colonisation. This was reflected, for instance, in the size of the peasant holding that required former plantation workers to work on the plantations for a few days for their subsistence. The assumption that Indians lived in a family context from the outset is curious because it was precisely during the immigration period that there was discussion about the significant shortage of women. Except for Fokken (2018), the incongruity of this assumption did not strike any of the later authors. Indeed, agronomists who regularly visited the field and should have known that there were few family households also departed from this assumption.

The debate among agronomists was not so much about the occurrence or non-occurrence of family households but rather the size of the plot. In the period of colonisation, it was a plot of 1 to 2 ha (Van den Houven van Oordt 1900), and in the 1920s, agronomists recommended 4 to 5 ha as a form of middle-class agriculture (Leys 1921). When it came to middle-class agriculture, most of those involved had Dutch farmers in mind rather than Girmitiyas (Brandon 1933; Memorie 1919; Verkade-Cartier van Dissel 1937). The 4 ha figure continued to circulate in administrative circles as a magic number in the 1920s until the 1950s. It remained central to post-World War II policy. Since 4 ha per family as a farm size was a desirable minimum size, it was soon referred to as the 'small farm'. Bürer (1956, p. 170) even speaks of 'a theory of the 4-hectare farm'. The definition of the 'small farmer' thus evolved from a farmer with 1 ha around the immigration period to a farmer with 4 ha in the mid-twentieth century.

In later research conducted by various agencies in Suriname, figures of 12 and even 16 ha were mentioned to make a family's living, but these were mechanised farms as discussed by Luning (1966, pp. 2–4). Later, Luning and Sital (1979, p. 179) counted plots of 24 ha among the mechanised farms. In all cases, the household was used as the basic unit, with incomes dependent on the cultivation of market crops, but without factoring in price developments. The increase in plot size alone, from 1 to 2 ha to 24 ha for a family's subsistence, was remarkable.

The 4 ha theory was tested for small rice farms in the 1950s by Bürer (1956, 1957). Using the results of the Subdivision Mechanisation of the Department of Agriculture and of the Prince Bernhard Polder (which was mainly a research centre and also conducted pilots) and the Wageningen Foundation, he distinguished three farm sizes of 1, 2 and 4 ha that had four degrees of mechanisation. The first was based entirely on manual labour; the second had only tillage with a tractor and a harrow, and light threshing by driving the tractor over the harvested stalks; on the third farm type, tillage was fully mechanised and harvesting partly mechanised, with the farmer using a small threshing machine; and the fourth type was fully mechanised, in which, in addition to the tractor and harrow, combines to harvest were used and the harvested paddy was transported by a tractor with a trailer. Bürer's observations and typology were derived from the experience in Nickerie (Bürer 1957, p. 196).

The amount of labour varied with farm type, area size and family size. Relying on an earlier budget survey (Budget Survey 1953), Bürer (1956) assumed that the average Hindostani family in the districts comprised 7.7 persons, about half of whom were aged 15 or older and thus could help with farming. An earlier analysis of farm type based entirely on manual labour had established that it took one person 310 hours to harvest 1 ha. Assuming 2 weeks at 8 h of labour per day, this meant that a one-person household could work 112 hours, or harvest 0.4 ha. In other words, a one-man rice farm could harvest a maximum of 0.4 ha in 14 days without the help of paid labour. It thus required at least three people to harvest 1 ha of paddy in a fortnight (Bürer 1956, p. 171).

Bürer (1957) compared this 'baseline measurement' with the three other farm types. As the degree of mechanisation increased, the number of manual labour hours decreased from 1160 hours per ha on the unmechanised farm to 300 hours on the fully mechanised farm. The number of machine hours on farm type 1 was 13, which increased to 21 hours on farm type 2 and decreased to 12 h on farm type 3. Bürer calculated the cost of labour, seed, ground rent and 5% storage loss at the different farm types and degrees of mechanisation and what the farms would yield at a price of 9 ct. per kg of paddy. The results are summarised in the table below.

Table 8.4 is disconcerting. On the most mechanised 4 ha farm, the net income was Sf 619 (last row, last column), and the income on the 4 ha farm type, which was based entirely on manual labour, was Sf 444. The incomes were set against an annual income of Sf 1000, which the Surinamese tax authorities considered a minimum taxable income and which Bürer (1957) defined as a 'reasonable income'. Vroon (1963, p. 31) mentions for the early years of 1950 that 'only 10.4% had a net annual income of more than Sf 1000'. Elsewhere in an extensive discussion of Suriname's economic problems he notes: 'From the summary and not always fully reliable data available, it was concluded that in 1951 the average gross annual

Table 8.4 Financial results in guilders at different rice farms, rounded down to whole guilders

Farm type	Size in ha	Gross yield	Expenditures in Sf	Net annual income in Sf
0	1	270	22	249
0	2	540	142	400
0	4	1080	636	444
1	1	270	90	180
1	2	540	278	262
1	4	1080	651	429
2	1	270	102	168
2	2	540	205	335
2	4	1080	442	638
3	1	270	115	155
3	2	540	231	310
3	4	1080	461	619

Source: Bürer (1957, p. 200)
0 = relies entirely on manual labour
1 = tillage and threshing mechanised
2 = tillage and threshing mechanised, harvesting mechanised
3 = all phases mechanised

income of the agricultural practitioner was barely Sf 350 compared to Sf 3750 and Sf 8750 of workers in industry and mining' (Vroon 1963, p. 40). Peasants thus earned on average less than one-tenth of the average income in industry and one-fifth of what mining brought in. These figures already indicate how widespread and deep the poverty in small farming must have been. Note that Vroon speaks of gross annual income and Bürer of taxable (i.e. net) income. The difference between gross and net incomes in smallholder agriculture was small because commercial inputs were mostly absent.

Bürer (1957, p. 199) pointed out that mechanisation resulted in time savings that could be exploited by growing other crops, but these incomes are not included in the above exercise. An unclear point in this quantification is that Bürer assumed three adult unpaid family members. This raises the question of whether each adult should have had a separate annual income. Even if one assumes that there were economies of scale through the sharing of fixed and other expenses in the household, the household income would have to be higher to accommodate three adults. Against this, it can be argued that the Sf 1000 standard most likely referred to a family that was totally cash dependent, while peasants mostly grow their own food. So their annual income is Sf 350 plus the homegrown food. Theoretically, that could be set at Sf 1000 (the taxable income), but that would mean that the cost of farmers' food would be Sf 1000—Sf 350 = Sf 650, i.e. 65% of a reasonable annual income. That would be an improbably high food quote. Even if we set the price of homegrown food at a generous 30% of taxable income (i.e. Sf 300 per year), the average income in the agricultural sector at Sf 350 + Sf 300 = Sf 650 would still contrast starkly with incomes in the other sectors.

From the Second Agricultural Census Suriname (TLS 1959, p. 58) it can be calculated that of the 13,800 rice farms, 10,145 (about three-quarters) had a size of 4 ha or less. Some of these farms also grew other crops such as maize, potato, cassava, groundnuts and peanuts, but the income from the main crop of rice, which incidentally also required the most investments, fell far short of what the Surinamese tax authorities had defined as a minimum taxable income (see Table 8.4). The conclusion to be drawn from the above is that poverty among farmers, even those who were best equipped for modern agriculture, was still widespread in the 1950s.

The poverty, given the wide disparity in sector incomes, must have affected peasant livelihoods down to the roots. Peasants near markets still had some outlets, but those who lived further from local markets had a considerably lower chance of survival. In other words, there was a spatial correlation between poverty and population concentration: the greater the distance from the concentration, the lower the chance of selling surpluses. But whether, and if so, how, the outlet opportunity was utilised was in turn dependent on infrastructure, health, household size and prices. It is therefore no exaggeration to say that life in agriculture was a drudgery. That the urbanisation of district residents continued even after World War II is not surprising: agriculture offered no livelihood prospects.

The preceding findings were based on developments in the district of Nickerie where the heart of mechanisation lay. The conditions for mechanisation (plot size, investments, knowledge, storage facilities, labour) were rarely present among smallholders. Van Dijk (1940, p. 107) had pointed this out, and a quarter of a century later, Luning (1966, p. 7) also made this clear. Assuming that mechanisation had progressed less in the old rice-growing areas (the districts of Saramacca, Commewijne and Suriname), it can be expected that here even more households were affected by poverty. It is therefore no coincidence that the migration from the districts of Suriname and Commewijne to Paramaribo, where the possibility existed of alternative employment and individual development, was greatest (Kruijer 1951; Van Dusseldorp 1963b).

Bürer's figures provide another reason for critical reflection. The budget survey on which he based labour needs consisted of 200 families in Paramaribo and 504 families in the districts. These families were required to take notes and fill in forms that were collected by inspectors every other day. However, the methodology indicated a bias in the sample in favour of families because single persons were omitted, while it was unclear whether incomplete households were also included in the sample. In other words, the smaller households that could push down the mean were omitted from the sample. Therefore, the availability of three to four working family members cannot have been an 'average' since the sample consisted only of families (Budget Survey 1953). It may be doubted whether the demographic balance had already recovered so soon after the Second World War and families on a large scale had sufficient labour at their disposal, let alone that so-called 'joint families' had emerged that could offer more labour (Adhin 1960; see also Lamur 1973).

Finally, three long-standing trends should be noted. First, many urban dwellers remained involved in agriculture. They were mostly descendants of villagers who had made a living in the city but had access to a piece of agricultural land as heirs or relatives. These people grew rice and vegetables for their own consumption. The second trend arose because of the long period between sowing and harvesting rice. That period enabled the peasant to earn additional income outside agriculture. This trend emerged only if there was some employment in the area of residence. And third, the more employment there was in the area, the harder the shift from agriculture to work outside. None of the three cases implied a permanent abandonment of agriculture.

Therefore, Luning (1966, p. 8) suggested that agrarian policy should be based on part-time farmers acquiring supplementary income outside agriculture. He also believed that farm size class could be better related to specialisation in specific products (rice, bananas, animal husbandry) that generated different incomes rather than focusing only on labour requirements.

8.5 Mechanisation

Issues of irrigation had been raised repeatedly in the past. In the late 1940s, they were officially raised by a committee commissioned by the Dutch government to investigate the possibilities of agricultural development in Suriname (Eysvoogel, Van Beukering and Verhoog 1948). The committee's central recommendation was the establishment of medium-sized farms with rice as the main crop in the great rainy season, followed by a shorter-growing second crop (most proposals concerned maize and soya). Implementation of the scheme obviously depended on soil conditions and needed infrastructure. More specifically, the need for irrigation and drainage was emphasised in order to achieve multiple harvests per year.

The idea of middle-class agriculture had been advocated earlier by various committees and individual authors, including Pyttersen (1922, pp. 30–31), who wrote: '[D]uring the war (he was referring to World War I, Author), a mechanical rice farm developed strongly in North America, which showed that under favourable conditions, a large percentage of livestock could be eliminated. The demands of such a business are such that cheap production is out of the question there. This is different in Suriname, where extremely favourable factors are in place. Nevertheless, practice will have to show whether the product can be cheaply produced there if machine labour is used. It is certain that rice cultivation in Suriname is very successful, as proven by the settlement. The Nickerie area in particular lends itself perfectly to this culture. There are a number of rice polders there, of which the Corantijn and Van Drimmelen polders are the most important. Here a good crop is produced without tillage.'

After Hewitt's first studies in the 1940s, the Surinamese government began, under the direction of Governor Kielstra, a policy of diversification and expansion. Hewitt's study showed that practically all farming families needed wage income from outside the farm for their subsistence. The proportion of this wage income varied by farm size: about 43% for cultivated areas of 1.2 ha or less, 30% for cultivated areas of 1.2 to 3.5 ha and 8% for cultivated areas of more than 4 ha (Schets

1951, pp. 22–23). In other words, the smaller the cultivated area of the farm, the more a supplementary wage income from outside the farm was required. This was a fundamental feature of small farming from the outset. In a later study, Hendriks and Joosten (1954, pp. 21–35) specified that this wage income varied not only with the size of the farm but also with the crops grown, which incidentally differed across ethnic groups.

As early as the 1920s, there were calls for larger-scale farms. The idea was to colonise Dutch farming families who would develop agriculture on larger pieces of land, and with more sophisticated means of production. The debate on middle-class farming continued in response to trial results of agricultural mechanisation in the 1930s, but especially in the second half of the 1940s (Brandon 1937, p. 302ff; Van Beukering 1952, p. 148ff; Verkade-Cartier van Dissel 1937, pp. 167–169).

The elements highlighted in this discussion were that:

(a) the surplus of agriculture for export remained too small to be of great significance;
(b) production could be significantly increased by occupying larger plots and using tractors since labour remained scarce in agriculture;
(c) transport costs were a disadvantage in international competition;
(d) the quality of rice needed to be improved in order to have any chance of increasing market share;
(e) the farm was of insufficient size to provide a full income to the farmer;
(f) agriculture needed to create sufficient employment to absorb some of the population growth.

Practically all these objectives can be found in modified or unmodified form in post-war agricultural planning (Kool 1964). The discussion on middle-class agriculture found a provisional end point in two initiatives. The first concerned the implementation of a colonisation project near the village of Wageningen in the Nickerie district. The initial proposal involved setting up a polder of about 5000 ha using development funds intended for the settlement of 70 Dutch farming families. For various reasons, this idea was abandoned and instead the initiative was continued as the Foundation for Machinal Agriculture (De Wit 1960, Ch. VII; Kool 1964, pp. 86–90). Linked to this was the secondary objective that the project should have a modernising impact on rice farmers in the vicinity. This function was assigned more specifically to the experimental Prins Bernhard Polder where the technical and economic issues of mechanisation in the tropics were investigated. The research function was also partly assigned to the Centre for Agricultural Research in Suriname (CELOS), which came under the Ministry of Agriculture, Livestock and Fisheries (LVV).

The second initiative involved an experiment with medium-sized farms on Lelydorp, a settlement near Paramaribo, and on plantation Slootwijk in Commewijne. At Lelydorp, the plan was to settle 35 farmers on plots of 5 to 11 ha, with a varied crop assortment and livestock, into viable farms. The project failed, not least because the estimated income was disappointing and aspiring farmers preferred to stay in the city (Kool 1964, pp. 83–86). Unlike the Lelydorp project, the Stichting Machinale Landbouw (SML), together with the Prins Bernhardpolder, were a success. The expansion of rice acreage in the Nickerie district must be largely attributed to these initiatives, especially to the Van Dijk concession, the SML and the Prins Bernhardpolder.

Mechanisation, described at this stage as the replacement of manual labour with machinery, was marked by the introduction of tractors. Tractor ploughing and harrowing resulted in time savings of 80% (Relyveld 1954, p. 80). Expansion of acreage was thereby facilitated, but the labour input during harvesting increased proportionally. The demand for labourers during harvesting could be curtailed by the use of mowers. The increase in paddy in turn placed demands on processing, drying and storage capacity. Consequently, the introduction of tractors was not an isolated phenomenon but also brought with it an increased demand for seasonal harvesters, combines, storage and transport facilities and paddy mills.

The use of machinery in agriculture was the subject of lengthy discussion in colonial circles. An official commission to investigate these opportunities was given to the engineering firm H.N. van Dijk in 1933, who reported back a few years later (Van Dijk 1940). From 1938, the Van Dijk firm and the Agricultural Extension Service demonstrated mechanical tillage and harvest processing to farmers in the Nickerie district (Relyveld 1954, p. 76ff.). By way of a 'hire service', the Van Dijk company was increasingly tilling the soil and harvesting and threshing paddy for farmers in the area. For further experimentation, this firm was allocated an area of 150 ha. This enabled it to become the largest private enterprise in the rice sector (Photo 8.2).

From 1943, the growth of the tractor increased rapidly. The table below is indicative of this process (Table 8.5).

Photo 8.2 Ir. H.N. van Dijk, pioneer of machine agriculture in Suriname, ca. 1927

Table 8.5 Tillage in ha and technology in Nickerie, 1943–1959

Soil tillage	1943	1947	1952	1959[a]
Planted area	3538	5600	6620	14,272
With cattle	1557	4000	2150	91
By handweeding	1757	1360	577	13
By machine	24	240	3893	14,980
Ownership of draft cattle	982	1600	2013	14,319
Ownership of tractors	4	20	128	409

Source: Relyveld (1954, p. 79)
[a]Data for 1959 taken from TLS (1959)

Table 8.6 Tillage of sawas in ha by district, 1959

District	Manual	With Oxes	Machines owned	Machines rented	Uncultivated
Nickerie					
0.48–20	6	92	1710	4691	9
> 20	7	–	8576	1	96
Saramacca					
0.48–20	7	8	92	807	1271
>20	2	1	169	121	85
Suriname					
0.48–20	243	4	290	3425	631
>20	10	2574	222	144	9
Commewijne					
0.48–20	72	244	20	705	1507
>20	2	11	46	41	39
Total land					
0.48–20	342	2924	2133	9766	3699
>20	14	38	9.019	310	141

Source: TLS (1959, p. 93)

Tractor use boomed in the 1950s. In the period 1943–1959, the amount of land tilled using draft cattle fell from 1557 to 91 ha, and that tilled by machine increased from 24 ha to just under 15,000. In 1959, the national total number of tractors was 622 (TLS 1959, p. 90), of which 409 were in the Nickerie district. Annex 8.2 shows that the concentration of oxen and ploughs in 1959 on the smaller farms ranged from 1 to 7 ha. Tractor ploughs, tractors and threshing machines were mainly found on farms larger than 7 ha.

The Second Agricultural Census (TLS 1959, p. 60) reported a total of 13,800 rice farms that together had devoted 28,546 ha to paddy. Of these, 2409 farms in Nickerie accounted for 15,191 ha, or 53% of the cultivated area. A comparison with the state of affairs in 1953 (see Table 8.3) shows that both the number of farms and the area under cultivation had declined. How the land was farmed and broken down by district and size class is shown in the condensed Table 8.6. The three non-agricultural districts have been omitted as well as the smallest farms (< 0.48 ha), while decimal figures have been rounded.

According to Annex 8.2, throughout Suriname in the late 1950s, 66% of tractors and 71% of threshing machines could be found on farms larger than 8 ha. The majority of tractor ploughs and harrows were also concentrated on farms larger than 8 ha. In contrast, 60% of plough oxen were found to be concentrated on farms of between 2 and 5 ha. The tendency to concentrate on advanced means of production on the larger farms is also reflected in the mode of sawa tilling (see Table 8.6 above). The table shows that in 1959, 95% of machined rice farms larger than 20 ha were located in the Nickerie district. In that district, the use of machinery was most advanced on both large and small farms, while in the district of Suriname, the smaller farms mainly rented the machinery. Given the difference between these two competing districts, the only possible conclusion is that by the late 1950s, the Nickerie district had taken over the lead in the use of modern technology.

This is not to say that Nickerie was free from agricultural issues. Firstly, it turned out to be very important that once the grains had matured, the rice variety had a strong culm to be harvested by machine. Skrivimankoti, the most commonly grown rice variety, did not possess this characteristic. The rice variety Rexora was a good alternative in many respects, but on the other hand, this crop grew slowly in the early days, so it was weak in fighting weeds. Effective mechanisation thus required the rice variety to be adapted as well. In other words, new rice varieties were necessary for successful mechanisation (Relyveld 1954, pp. 82–83).

In addition, the movement of threshing machines proved time-consuming due to their large size, while passage over normal plot bridges was sometimes impossible. The fact that a large number of operations took place during one (short) period of the year led to underutilisation of the new machinery. After tilling the land, the tractor could possibly still be used for transport and threshing. By renting it out, one could increase the profitability of these investments. But because the land bore only one crop per year, the tractor's productive period was limited to one season. This underutilisation gave rise to attempts to grow two production cycles or another second crop per year. Skrivimankoti would have to be replaced as this variety was not suitable as a second crop in the same year.

More intensive cultivation of the land would also affect the natural productivity of the soil. Conservation and the increase in soil productivity took place by flooding the fields for a certain time. However, it meant that when two crops (or one crop twice a year) were grown, not only did drainage and water run-off have to be adequate but 'humus preservation' also had to be done in some other way (Eysvoogel 1950, pp. 350–353; Introductory 1953, p. 5). That is to say, the use of fertilisers was unavoidable.

There were also doubts regarding the family as the unit of the farm. Proponents of a large-scale farm set-up were fiscally oriented and pointed out that the family farm would contribute little to the exchequer or to national prosperity (Van Beukering 1952, p. 149). In addition to favouring a large-scale farm set-up, proponents argued that the low enthusiasm for working in agriculture and the family as an economic unit made it vulnerable in times of illness of one or more family members. Moreover, it turned out that the required investments could by no means all be borne by a family. For these reasons, some experts suggested that the purchase of the equipment should be done in a cooperative way so that the burden could be spread and a fee could be paid in the case of outage (Eysvoogel 1950, p. 352).

The traditional ethnic cooperative set-up gradually disappeared. Peasants had become commercial and specialised farmers and competitors of each other. Attempts to establish cooperatives were generally most successful in Nickerie. But cooperation no longer consisted of helping during the production process (sowing, harvesting, borrowing planting materials and the like) but of setting up infrastructural works such as building roads, reclaiming land and buying a rice paddy mill (Hewitt 1941; Kruijer 1949; Relyveld 1954).

The photograph below shows a rice paddy miller who was not exactly in good spirits, as also can be gauged from the shabby building in the background. Both things, the appearance of the miller and the building, indicate the level of prosperity in Nickerie, which was the most prosperous agricultural district in 1947 (Photo 8.3).

In 1953, the government decided to set aside part of rice exports for agricultural cooperatives and associations. Such attempts to set up cooperatives were also made in other districts, but they were unsuccessful. The failure of these attempts was blamed not so much on economic forces such as competition as on mutual envy (Kruijer 1949, pp. 214, 218ff). Cooperatives now took shape in formal organisations and were no longer ethnic cooperatives. To this can be added the fact that employment outside agriculture increased, which was also at the expense of cooperation.

Photo 8.3 Owner of a rice paddy mill in Nickerie, 1947

8.6 The Rice Issue

Increased land productivity, the issuance of larger plots of land, mechanisation, the introduction of new rice varieties and infrastructure improvements accelerated rice production in the Nickerie district. Production in that district grew more than three-fold in the period 1945–1960, whereas production in the Suriname district, while more than doubling, grew at a slower rate. However, all districts saw an increase in production (see Table 8.7). In the period 1950–1958, the production of large-scale agriculture, comprising the paddy production of Wageningen, Prins Bernhardpolder and Van Dijk's concession, grew from 338,000 kg to 22,258,000 kg. The big jumps in Nickerian production took place in the years 1953/1954, 1955/1956 and 1957/1958. The table shows that the average yield per ha fluctuated but was generally lower (and sometimes considerably lower) than the 3000 kg per ha as assumed for a long time. This was also true for the production of large-scale farming (Abrahamse 1959, p. 52). One possible explanation is that the experiments produced lower yields.

Storage represented an additional, though not insignificant, issue. A rapporteur of the Food and Agriculture Organization described the problem as follows: '[T]he storage of paddy on farms is very insufficient. The paddy is usually stored in improvised sheds placed outside the houses or under them when they are built on piles. As a result of this there is much deterioration of the product as well as rat and insect damage' (De Vos 1956a, p. 6; see also Caribbean 1955, p. 25). Thus, for the period

Table 8.7 Production (in 1000 kg) and area under paddy, including second harvest, by rice district, 1945–1960

Year	Nickerie		Saramacca		Suriname		Commewijne	
	Production	Size	Production	Size	Production	Size	Production	Size
1945	13,730	4450	6118	1887	8539	5957	3835	1924
1946	16,259	5150	6717	1761	20,491	7118	6419	2397
1947	13,977	5600	6043	2249	13,598	6907	4334	2661
1948	16,800	5232	7800	2375	23,700	7537	7887	2643
1949	16,500	5335	7470	2093	18,600	7537	6579	2190
1950	17,920	5830	6412	1959	18,832	7311	6181	2344
1951	19,000	6100	7340	2219	23,328	7684	6778	2400
1952	16,120	6620	8362	2510	21,495	7847	6603	2357
1953	19,973	6693	6878	2393	20,387	7581	8524	3001
1954	25,006	7638	8197	2521	23,393	8878	8584	2854
1955	26,460	8830	7136	2224	22,307	8300	7417	2501
1956	32,624	11,177	7302	2251	22,564	8646	7500	2525
1957	36,517	14,171	3862	2203	10,867	9023	3323	2590
1958	44,638	15,203	7826	2754	22,972	9495	8074	3268
1959	42,568	15,191	7182	2567	20,406	7672	7625	2767
1960	46,574	16,718	7919	2571	18,512	8007	5692	2352

Source: 1945–1958, Abrahamse, 1959, pp. 49–50 and DSL for 1960

1945–1957, it was estimated that 15% of the production was lost to insect damage (Abrahamse 1959, p. 55).

The problems of paddy storage consisted of protection from moisture, insects and rodents and theft, and filling and emptying the silos. Especially for the latter, silos had to meet specific requirements for materials and construction, insulation, wall pressure and roofs (De Vos 1956b). As the period between purchase and delivery lengthened, greater loss of quality and weight occurred. However, since the producer-seller bore the entire risk, they were forced to supplement batches reduced in quantity and quality with additional purchases. Profits could thus turn into losses. The same problem occurred after the batches were released for export. These too had to be replenished, while the reduced quality weakened the competitive position.

What about rice sales? The small domestic market had long been saturated. Future domestic disposal of the rice surplus depended on population and income growth. An economist of the time estimated the increase in domestic sales for the period 1955–1965, including the effect of rising individual purchasing power, at around 2–2.5% per annum. Thus, in the 10 years between 1955 and 1965, the sales increased by up to 25%. The increase also required the provision of better planting materials, controlling of insect pests and prevention of spoilage at harvest and in storage (Kool 1955, p. 52). The agricultural policy for that period envisaged a tripling of the agricultural area in 10 years with the aim of producing a greater rice surplus for export.

This possibility was considered favourably. In 1952, the Netherlands Economic Institute (NEI) was commissioned to take a closer look at the sales potential of Surinamese rice on the foreign market until 1965. Based on the expectation that production in Asia would not be able to fully meet the future demand for rice, it was considered possible that 'the outlets for Surinamese rice – if not priced too high – are favourable' (NEI 1952, p. 16). This study hardly addressed the quality of the product; the increase in Suriname's market share was quantified based on the expectation that the percentage growth in demand for rice would exceed supply. The issues of storage, transportation and quality were also not included in the analysis. The optimism of the authors of the NEI report was in line with the estimation of the Dutch Ministry of Economic Affairs. According to that department, the emerging large-scale agriculture in Suriname, and in particular the Wageningen project, could supply the growing demand in the Netherlands with its rice surplus (De Wit 1960, pp. 60–63).

However, exports were subject to restrictions. During the war years, production was bought up by the government to secure domestic needs for rice. After that, an annual paddy decree was promulgated in which sale, delivery and issuance were forbidden without government permission. The exception was farmers in the Paramaribo area, who were free to deliver their produce to the city. From 1946, exports of paddy and rice were regulated by the government. The practice was that the governor, through the district commissioners, bought up paddy at a maximum price. Inspection was carried out by officials of the Department of Agriculture. This method sought to protect urban consumers from excessively high prices and also guarantee a certain quality of rice (Elenbaas 1952, pp. 2–11; Van As and del Prado

1960, p. 8). Since the government did not have its own silos, the farmer was obliged to keep the stock for a fee until delivery. Any profits made by the government were distributed to farmers in the form of bonuses. After meeting domestic needs, the surplus was released for export.

This practice had important implications for the quality and quantity of rice and the pace of marketing. The Suriname Business Association (VSB) noted as its main objections: 'a) on the part of producers that they receive too low a payment for their paddy, or at least do not make a price related to the world market price. The bonus fund aimed at compensating them for that difference does not work satisfactorily; b) on the part of traders that they have to wait and see whether the government offers paddy rice for sale; this makes real trade with foreign countries impossible' (Elenbaas 1952, p. 4). It was added that the vicinity of Paramaribo could negotiate the market price for their product.

This situation set interest groups in motion. In 1952, the National Association Nickerie Vooruit (NVNV) addressed an open letter to the Minister of Economic Affairs. It highlighted the growing need to capture permanent outlets on the world market. The NVNV pointed to the rise of the United States as a rice producer and the restrictive laws and regulations in Suriname. Its urgent advice was to leave buy-back free in order to guarantee the grower the highest price for the best quality, in addition to imposing quality requirements on exporters (Van Dijk 1952, pp. 21–25).

In 1955, the buy-back system was revised and a limited number of private individuals were allowed to export directly. The sharp increase in export (Annex 8.1) can be associated with this revision, but the persistently higher price on the foreign market is also notable. The freer foreign rice trade is reflected in the convergence of the domestic and export price of rice seen since 1955. According to Annex 8.1, the quantity of rice exported doubled from 1954 to 1955 (and then declined), but this was offset by a fall in the average export price. It is unlikely that the fall in prices affected the decline in export, for both the rise in export volume and the fall in export price continued until 1960.

8.7 Conclusions

What did the mechanisation of rice farming mean for Hindostani farmers? Motivated by economic and planning motives (export promotion, income for the government and the colonisation of Dutch farmers), the Dutch government pursued a twin-track policy after the Second World War: one focused on the mechanisation of rice farming and the other on countering urbanisation by improving the lived environment in the districts.

To achieve the two goals of mechanisation and curbing urbanisation, the government implemented some reforms after World War II. These related to land allocation, the creation of new polders, the initiation of experimental agriculture (including by the Van Dijk firm, the Prins Bernhardpolder, the Wageningen Foundation, CELOS and the Ministry of Agriculture, Livestock and Fisheries), the promotion of mechanised family farms, and the improvement of production conditions and marketing. In the decades after World War II, these efforts proved to be in vain. Urbanisation continued and life in the districts remained meagre, mainly due to poverty and a lack of further education. Moreover, prospects of a good life in the districts were lacking and failing. It is therefore no coincidence that communalisation did not take off in the various district centres.

In 1953, more than half of the farmers owned plots smaller than 2 ha and 78% of them had plots smaller than 4 ha. Not only was the size of their plots a hard indicator of poverty, but the quality of the land, its location in relation to markets, health, having cooperating relatives, ownership of labour tools and animals and, above all, employment opportunities in the neighbourhood were all factors that determined household income. Due to the policy focus on production to promote exports and increase state revenues, this issue of 'persistent poverty'—probably the most important characteristic of the Hindostani farmer—has hardly been addressed. What did receive scholarly and political attention were the successes in Nickerie, which incidentally may be put into perspective. There, it was a few fortunate farmers who made it to some prosperity.

What do these successes mean for small mechanised farms and high land productivity, i.e. the inverse relationship? Actually, not very much. Because of the mixed nature of Surinamese farms and because of the large fallow period, it is impossible to determine whether the land on the small farms was intensively cultivated and whether the yield per ha was on average higher than on the larger farms. Moreover, two other data contradict the emergence of an inverse relationship. Many former residents of the plots left the districts or took other jobs near where they lived. Farming then became a secondary occupation, either to earn additional income or to produce food crops for their own consumption. This situation renders any comparison of full-time farmers with farmers who possessed no or few alternative options invalid. And perhaps more importantly, the traditional Girmitiya peasant was transformed into a commercial farmer that had to specialise in rice as a commodity. While there was a strong government-induced—tendency towards commercialisation, this could hardly be considered the beginning of a capitalist development.

What mainly emerged in the Nickerie district were mechanised family farms that could operate more independently than those in other districts. The small businesses were predominantly family enterprises and hence the use of wage labour was minimal. Capitalist development presupposes a labour force and expansion of acreage. Capitalist rice farming presupposes greater land ownership, less intensive land use, rotation of land use and thus on average a lower yield per ha of land. These conditions were hardly met.

In the period under review, two turning points became visible in Surinamese rice farming. The first consisted of the spread of machine farming, especially in the Nickerie district. This changed the character of Hindostani farming. Mechanisation transformed the smallholder farmer into a modern farmer and changed the face of Surinamese rice farming. The second development was urbanisation. This development was not always an absolute break with agriculture but rather had various degrees of transition. Sometimes it was family members moving to the city while the family remained in the district. And sometimes it was entire families who moved to the city but mostly kept a piece of land to grow their own living. This development also changed the character of Surinamese rice farming.

The two typical developments among rice farmers, namely the rise of mechanised family farms and part-time farming, gave a decisive turn to relations within the ethnic community. The inter-ethnic cooperation between farmers, especially of the first generation, expressed in deforestation, house building, land and landscape layout, sowing, harvesting, transporting and the like, declined. Families became more independent with mechanisation, but collective action, such as road construction or maintenance, remained necessary but decreased. However, ethnic ties and cooperation as a moral community continued in the social and religious spheres. That ethnic loyalty was foundational for the Girmitiya community. It gave rise to a new dimension of ethnic life when the Hindostanis established a political party and became a political community.

References

Abrahamse J (1959) Padi: Verslag van de in 1958 Gehouden Opnamen Betreffende de Oppervlakte en de Produktie. Ministerie van LVV, Paramaribo

Adhin JH (1960) Over de 'joint family' der Hindostanen. Nieuwe West-Indische Gids 40(1):17–27

Adhin JH (1961) Development planning in Surinam in historical perspective. Stenfert Kroese, Leiden

Akram-Lodhi H (2001) Vietnam's agriculture: Is there an inverse relationship? Den Haag: ISS Working Papers Series no. 348

Banking 1948 Rapport Betreffende het Bankwezen in Suriname. No place mentioned

Beck S (1924) Landbouwbedrijfsvorm en landbouwopleiding. In: Beck S (ed) Surinaamsche Problemen. Kersten, Paramaribo, pp 1–44

Brandon HG (1933) Proefneming voor Machinale Hoeverijstculuur op de plantage Guadeloupe in Suriname in 1925–1926. West-Indische Gids 14(1):81–98

Brandon HG (1937) De Proefneming van de Surinaamse Hoeve Maatschappij. West-Indische Gids 18(1):293–306

Budget Survey (1953) Budgetonderzoek gehouden in Suriname in Februari–Maart 1952. Stichting Planbureau, Paramaribo

Bürer T (1956) Het Niet–Gemechaniseerde Rijstbedrijf van 4 ha. De Surinaamse Landbouw 4(5):170–174

Bürer T (1957) Mechanisatie van het Kleine Rijstbedrijf. De Surinaamse Landbouw 5(5/6):196–202

Caribbean (1955) Caribbean market survey, Rice. Port of Spain, Caribbean Commission

Courbois OA (1958) The financing of agriculture in Surinam. Caribbean Commission, Trinidad

De Vos L (1956a) Storage and processing of rice in Suriname. Ministerie van LVV, Paramaribo

De Vos L (1956b) De opslag van padie in Suriname. De Surinaamse Landbouw, jrg 4(5):181–192

De Wit TPM (1960) The Wageningen rice project: a study on the development of a mechanized rice farming in the wet Tropics. Doctoral dissertation, Wageningen

Djurfeldt G (1984) Classical discussions of capital and peasantry: A critique. In: Harris J (ed) Rural development: Theories of peasant economy and agrarian change. Hutchinson & Co, London, pp 139–159

Dorward A (1999) Farm size and productivity in Malawian smallholder agriculture. J Dev Stud 35(5):141–161

Dyer G (1991) Farm size – farm productivity re–examined: Evidence from rural Egypt. J Peasant Stud 19(1):59–92

Elenbaas GM (1952) Het Padiebesluit: Memorandum van de Voorzitter van het Surinaams Bedrijfsleven. Vereniging Surinaams Bedrijfsleven, Paramaribo

Eswin (1956) Jaaruitgave van de Economische Stichting West Indië-Nederland, 's-Gravenhage

Eysvoogel WF (1950) Mechanisatie van de Surinaamse Landbouw. Economisch-Statistische Berichten 1721(3 mei):350–353

Fokken M (2018) Beyond being koelies and kantráki. Constructing Hindostani identities in Suriname in the era of indenture 1873–1921. Doctoral dissertation, University of Groningen, Groningen

Gaurav S, Mishra S (2015) Farm size and returns to cultivation in India: Revisiting an old debate. Oxf Dev Stud 44(2):165–193

Gowricharn RS (1991) Economische transformatie en de staat. Over agrarische modernisering en economische ontwikkeling in Suriname, 1930–1960. Doctoral dissertation, University of Utrecht

Hendriks JAH, Joosten WJM (1954) De Resultaten van een landbouw–proeftelling in de omgeving van Lelydorp. De Surinaamse Landbouw 2(1):21–35

Hewitt WE (1941) De sociaal-economische toestand van klein-landbouwers aan de Boven-Commewijne. Jaarverslag van het Departement van Landbouw-Economische Zaken, Vol. not reported, p 73–87

Introductory (1953) Introductory report on mechanizing agriculture in Suriname, Report no. 1. Wageningen/Holland, Netherlands Advisory Board for the Mechanization of Agriculture in the Overseas Territories

Kool R (1955) De toekomstige binnenlandse afzet van de Surinaamse landbouw. De Surinaamse Landbouw 3(2):48–53

Kool R (1964) Agricultural planning in Suriname 1950–1960: An evaluation. Wageningen, Veenman & Zonen

Kruijer G (1949) Landbouw-Cooperatie onder de Aziatische bevolkingsgroepen van Suriname en Brits-Guyana. Nieuwe West-Indische Gids 30(1):209–235

Kruijer G (1951) Urbanisme in Suriname. Tijdschrift van het Koninklijk Nederlands Aardrijkskundig Genootschap, jrg 58:31–63

Lamur H (1973) The demographic evolution of Surinam 1920–1970: A socio-demographic analysis. Martinus Nijhoff, The Hague

Leys JJ (1921) De Kleine Landbouwers en het Klein Landbouwbedrijf. West-Indië, Landbouwkundig Tijdschrift voor Suriname en Curaçao 6(1):31–44

Luning H (1966) Bedrijfsgrootte, werkgelegenheid en agrarische planning. De Surinaamse Landbouw 14(1&2):36–44

Luning H (1967) Economic aspects of low labour-income farming. Doctoral dissertation, University Wageningen

Luning H, Sital P (1979) The economic transformation of small holder rice farming in Surinam. In: Cross M, Marks A (eds) Peasants, plantations and rural communities in the Caribbean. Leiden, KITLV, pp 193–219

Memorie (1919) Memorie van het Comité tot Bevordering van den Middenstandslandbouw in Suriname, op 29 augustus 1919 den Gouverneur Aangeboden. West-Indische Gids 1(2):163–176

Morenc J (1988) Surinaamse Kleine Landbouw en Landbouwbeleid: Een Structurele Analyse. Doctoral dissertation, Radboud University of Nijmegen

NEI (1952) Onderzoek Naar de Afzetmogelijkheden van Surinaamse Rijst. Nederlands Economisch Instituut, Rotterdam

Patnaik U (1976) Class differentiation within the peasantry: an approach to analysis of Indian Agriculture. Econ Polit Wkly 39:82–101

Pyttersen T (1922) Rapport omtrent de uitkomsten van een voorloopig onderzoek naar de levensvatbaarheid van landbouwersbedrijven in Suriname met toepassing van moderne cultuurmethodes. s–Landsdrukkerij, s-Gravenhage

Relyveld WL (1954) Mechanisatie van de landbouw in Nickerie. De Surinaamse Landbouw 2(2):75–84
Roy P (1981) Transition in agriculture: empirical indicators and results (evidence from Punjab, India). J Peasant Stud 8(2):213–241
Schets (1951) Eerste Schets van de Ontwikkelingsmogelijkheden van Suriname, deel ll, De Plannen. Paramaribo/Den Haag
TLS (1959) Tweede Landbouwtelling Suriname, november 1959. Ministerie van LVV, in samenwerking met ABS, Paramaribo
Van As JB, Del Prado H (1960) Nota Inzake Garantieprijzen voor Landbouwproducten. Ministerie van LVV, Paramaribo
Van Beukering JA (1952) Types of farming. Caribbean Economic Review IV(1&2):140–163
Van Dijk HN (1940) Mechanische Rijstcultuur: Economische en Technische Resultaten Na Zes Jaar Pioniersarbeid te Nieuw–Nickerie, Suriname. Amsterdam: Paris
Van Dijk HN (1952) Suriname's rijstcultuur wordt bedreigd: Open brief aan de Landsminister te Paramaribo. Paramaribo, Nationale Vereniging 'Nickerie Vooruit'. VSB, Paramaribo
Van Dusseldorp DBWM (1963a) Het Centraplan voor Suriname: Some aspects of town and country planning in a developing country. Wageningen, Veenman & Zonen
Van Dusseldorp DBWM (1963b) Geografische Mobiliteit en de ontwikkeling van Suriname. Bijdragen tot de Taal-, Land-, en Volkenkunde 199(1):18–55
Van Dusseldorp DBWM (1967) Meerdimensionale Overheidsplanning. De overheidsplanning van Suriname in de periode 1952–1964, getoetst aan een relatiemodel. Doctoral dissertation, Wageningen
Van Gelder P (1984) Werken Onder een Boom; Dynamiek en Informele Sector: de Situatie in Groot-Paramaribo, Suriname. Doctoral dissertation, University of Amsterdam, Amsterdam
Verkade-Cartier van Dissel EF (1937) De mogelijkheid van landbouwkolonisatie van blanken in Suriname. Doctoral dissertation, University of Utrecht
Vroon IJ (1963) Voorgeschiedenis, opzet en resultaten van het Surinaamse tienjarenplan. Nieuwe West-Indische Gids 43(1):25–74
Welvaartsfonds (1955) Samenvattend Eindverslag van het Welvaartsfonds Suriname. Tevens Verslag over het Jaar 1954, Paramaribo

Chapter 9
Ethnic and Economic Politics, 1945–1980

Abstract This chapter deals with the ethnic base of the VHP, the ethnic Hindostani political party. It argues that peasants' interests were hardly represented by the VHP whose leadership had an urban orientation. This neglect of the agrarian constituency led to rifts in the political community. Ethnic loyalties were differentiated, first by regions (city and districts), and later by classes and political views. Nevertheless, the Hindostani community required some ethnic unity to bear the VHP. This argument is structured as follows. Section 9.2 discusses some elements of the unification of the Hindostani community. Section 9.3 examines the background of the VHP, while Sect. 9.4 analyses the socio-economic planning in Suriname after WW II. Section 9.5 examines the link between the agrarian background of the VHP and agrarian planning in post-war Suriname, in an attempt to establish whether the party defended the interests of the Girmitiya peasantry. Section 9.6 addresses the internal differentiation of loyalties or political oppositions within the Hindostani community, specifically the separation between economic and moral community and the spatial division between district and city.

Keywords Ethnic cohesion · Urban orientation · Agrarian planning · Ethnic party · Ethnic loyalties

9.1 Introduction

The period after World War II saw two major changes. The first has already been touched upon and concerned the mechanisation of rice farming, which was mainly concentrated in the Nickerie district. The second was that the ethnic cohesion of the Hindostani population group changed in character. The mechanisation made families in middle-class agriculture less dependent on neighbours and relatives for sowing, harvesting and transporting. The bonding effect of the agrarian human capital had diminished. However, this increased self-reliance was partly based on income earned outside agriculture. As a result, the economic and cultural spheres were partially separated. Cultural affiliation continued in the social sphere, for instance in

R. Gowricharn, *The Girmitiya Peasants in Suriname*,
https://doi.org/10.1007/978-3-031-67961-2_9

marriages, celebration of religious or other festivals and political allegiances. The ethnic and moral community still operated.

The second change was the shift in ethnic loyalties from the family level to the community level and thence to the political level. This shift was reflected and encouraged by the formation of political parties through which Hindostanis also manifested themselves as a political community. After World War II the Hindostani community faced the task of uniting politically, which, above all, meant transcending religious differences. Some authors characterise the United Hindostani Party (VHP) as a party of traders (Hoppe 1976), but the distinctive feature of this party, especially in the period of its formation, was the ethnic loyalty of its agrarian constituency. To keep this loyalty intact—that is, to survive as a peasant party—the VHP had to propagate a unified image, which consisted mainly of a chauvinistic ideology and a homogeneous image of Hindostani history (see essays in Azimullah et al. 1963; De Klerk 1953). To this end, Hindostani ideologists had to propagate a material and immaterial culture.

Such creation of a culture is known in the literature as an 'invention of tradition' (Hobsbawm and Ranger 1983). The concept has mainly been used to refer to political mobilisation in the context of nationalism. Traditions that fell into oblivion or did not exist at all were propagated from a nationalist-inspired image of the past to give the nation an identity or past, heroic or otherwise. In line with this phenomenon, Sollors (1989) launched the concept of 'invention of ethnicity' to denote ethnic ties and institutions that may or may not exist. These terms have nationalistic or chauvinistic overtones. It mainly meant creating a specific image of history, a cultural image of gender relations (especially the female image), of cultural traditions and of a homogeneous ethnic group.

Surinamese Girmitiyas' ethnic unity at the macro level and its mobilisation by the VHP raises the question of whether the party looked after the economic interests of the peasants. Put another way, did the shift of ethnic loyalties from the household level to the macro level lead to a policy in favour of Hindostani peasants? The question inquires about a specific agrarian policy, something that in the colonial period was entirely in the hands of the Surinamese colonial elite or of the Netherlands. Now that in post-war Suriname, Hindostanis owned their own political party, the question arises as to whether the Girmitiya political elite intervened on behalf of the peasants. This chapter thus focuses on whether the shift from ethnic loyalties translated into agrarian advocacy.

The chapter deals with the ethnic base of the VHP and the representation of the peasants' interests. It argues that peasants' interests were substantially represented only in Nickerie, although it was insufficient to achieve a sustainable flourishing rice culture. Despite the establishment of middle-class farms, families continued to struggle to make a decent living because most of the government's attention was focused on the Wageningen project and the Prins Bernhard Experimental Polder, both of which were mainly geared towards rice exports. This neglect of the agrarian constituency—despite the invention of ethnicity—led to rifts in the political community. Ethnic loyalties were distinguished, first by regions (city and districts), later by classes and political views. This was reflected in the fact that the VHP had as many as three political splits even before Suriname's independence in 1975 (Dew 1978).

In the next section, I discuss some elements of the unification of the Hindostani community. Section 9.3 examines the background of the VHP, a heterogeneous ethnic party that contained few recognised but important contradictions. In Sect. 9.4, the socio-economic planning in Suriname is analysed as far as is relevant for agriculturalists, exploring whether Girmitiya peasants benefited from the new economic policy. Section 9.5 examines the link between the agrarian background of the VHP and agrarian planning in post-war Suriname, in an attempt to establish whether the party defended the interests of the Girmitiya peasantry. Section 9.6 addresses the internal differentiation of loyalties or political oppositions within the Hindostani community, specifically the separation between economic and moral community and the spatial division between district and city. The conclusion evaluates the findings of this chapter for Suriname's rice politics.

9.2 Ethnic Cohesion

The experience on the plantations was diverse as it involved several categories in regard to people (gender, religion, age and health). This diversity also applied to the peasants. Moreover, as peasants, they practised a variety of occupations, such as fishermen, paddy mill owners, transporters, barbers, shopkeepers, gold- and silversmiths and traders whose activities, incidentally, were not always mutually exclusive. The shopkeeper, for instance, also planted rice or other horticultural and agricultural crops and kept poultry; the peasants and related professional tradesmen had similar experiences or knew about them from neighbours or relatives. This led to the cultivation of more or less the same crops or to the development of the same pattern of living: despite the diversity of occupations, they had the same hustle and bustle in a given period and the same 'free time'. This was the case in British India (Yang 1979) and it was no different in Suriname (Bürer 1956).

This similarity in crop cultivation also led to the consumption of the same 'ethnic food'. Rice had already become a staple food since World War I, but numerous other elements in the diet remained specifically Hindostani. Chapter 6 pointed out the continuity in the use of vegetables, fruits, herbs, medicinal plants and flowers between British India and Suriname. Hindostani food in Suriname was characterised by the use of different types of peas like the dáhl (especially the urdi and yellow peas) and the karhi (a thicker form of the dáhl made from yellow peas), the use of chutney (a type of sambal) made from fruits, pepper, oil, garlic, onions and salt, and masálá, which enabled various curry dishes (both vegetarian and with fish and meat), the roti, and a wide variety of snacks of which the phulauri, sehená and ghughari (fried chickpeas) were and still are the most in vogue. This continuity reflects the fact that culinary skills and tastes remained largely intact, i.e. that deskilling in this domain had been slowed down considerably.

Food thus became a feature of Hindostani identity. Before WWII, not all identity-forming institutions were present or fully developed. One such institution was the family. The relationship between the sexes, however, was slowly recovering. In addition, a system developed in which a pandit (Hindu priest) was the focal point in a network of households that he served, a system known as the 'jajmáni' system (Bihari 1974). In an interview (dated 21 April 2011), former president Nannan Panday of the Sanatan Dharm, the largest Hindu organisation in Suriname, recounted that pandits helped forge ethnic group cohesion. 'These men were not only the links between families, but they also taught their fellow group members about the culture they descended from. They, so to speak, wove the community together. The same role was played by the Muslim religious leaders.'

The development of a community language, Sarnami Hindostani (literally, Surinamese Hindostani), was a peculiarity because it did not take off in most Caribbean diaspora countries. This language, said to be an amalgamation of Awadhi, Bhojpuri and some other British Indian languages, incorporated words from Dutch and Sranan. But it has an ambiguous status and can be put down as the lingua franca of Hindostanis, as an ethnic language, as a dialect or as a language in its own right (Gobardhan 2001, pp. 426–431). Sarnami varies greatly from generation to generation in terms of idiom and also has wide variations depending on whether the speaker is from the city or a district.

In addition, Sarnami in Nickerie has a specific variant due to Guyanese influences. Until recently, this language was the preferred means of communication within the Hindostani community, but it is losing ground due to the rise of Sranantongo, originally the lingua franca of Creoles. Whether Sarnami has contributed to the cohesion of Hindostanis is unclear because Hindostani communities in Guyana and Trinidad, for example, have not developed their own language and yet are considered an ethnic group. It is plausible, however, that having one's own language increases recognition and acceptance among peer groups in a diaspora.

Sarnami has always held a subordinate position to Hindi. On official occasions such as conversations on the radio, speeches by politicians and dignitaries, and rituals performed by pandits, speakers prefer to deploy Hindi. The issue is not whether Hindi is spoken correctly (according to Indians, it rarely is), but rather the higher status attached to it. The use of Hindi creates the suggestion that the speaker has access to Indian sources, especially the written word such as religious scriptures, has mastered the spoken word such as in Indian films and understands songs. Hindi speakers, by using this language, demonstrate their culturally superior status to Sarnami speakers, their authenticity and their level of development. De facto, Hindi was the undisputed language of the Hindostani elite until the 1980s.

Concomitantly, the community developed a corpus that was claimed to have historical significance. Most of this part is oral in nature and should be qualified as the cultural heritage. Various cultural expressions (the song and dance known as the *baithak gáná*, *londá ke nác*, *birahá* and *ahirwá ke nác*) developed in and through Sarnami. This inherited folk culture retained its original religious nature but was strongly determined in its further development by the modernisation of musical instruments and acculturation. In addition, Sarnami functioned as a vehicle of 'oral literature' and songs in which the sufferings and sorrows of the contract workers, especially deception, miserable crossings and hard plantation life, were sung about.

Peasants hardly feature in these texts. The significance of Sarnami should not be overestimated. Elahi (2018) argues, using YouTube videos, that related song and dance forms, variations aside, also occur in other Indian diaspora countries. She described the evolution of musical instruments associated with faster rhythms, which made dancing more contemporary and exuberant. Since similar forms of singing in local English were developed in neighbouring Guyana and in Trinidad, the conclusion is obvious that technology was more important to song, music and dance culture than Sarnami.

Arya (1968) collected various songs among Surinamese Hindostanis in the late 1950s. Most of his study consists of carefully classified songs and their translations. According to him, the songs belonged to ritual, work, life cycle, annual cycle, castes and labour, with a category of 'other songs' as the final section. With the exception of 'other songs', he said the songs could be traced back to traditions in eastern Uttar Pradesh and Bihar. The 'other songs' category is interesting because it reflected the grief and loss of the British Indian immigrants. Arya's discussion of this category centres on the *birahá*, which De Klerk (1953, p. 70) described as 'plaintive love songs'. In the collection of *birahá* songs, for which Arya visited several districts in the late 1950s, there is not a single song in which peasants' emotions feature. Sometimes there is a phrase with an allusion to agriculture, but its origin may just as well have been in British India. There are songs related to Suriname and India, but they only depict a bond without any reference to grief and loss (Arya 1968, pp. 162–170).

Although it is likely that grief and loss as well as nostalgia and homesickness marked the lives of the British Indians (see Legêne 1936) and these emotions translated into songs (and that Arya overlooked them), it is most likely that the lyrics about these emotions were written by a later generation. Currently, it is part of the *communis opinio* to interpret texts of songs as descriptions and explanations of the Girmitya experience. The songs may also have been composed because the melodies or rhythms are catchy, because they are easy to dance to, to tease others, because the lyrics rhyme or simply for entertainment and fun. The tribulations perceived by later generations of the British Indian indentured labourers created a heritage and contributed to the ethnic cohesion of Hindostanis. Such is the social and political contribution of oral culture.

Celebrations also had binding functions in the community as they promoted ethnic identification, internal social networking and the transmission of culture. This includes weddings (*biyás*), *kathás* and *hawans* (religious services, mostly held at home) and other religious or community festivals such as Diwáli, Holi, Ramlilá, Eid and Eid-ul-fitr. The festivals, like the songs, were later modernised and adapted to local conditions. For instance, wedding rituals that lasted a whole day in the 1940s were reduced to 1 or 2 h after the turn of the millennium, the Diwáli celebration was also reduced to a day (while in parts of India it is celebrated for a week), and the religious play Ramlilá, which was performed annually in meadows or on football pitches, declined in significance due to urbanisation. The British Indians had more holidays, as evidenced by the large number of days off for religious practitioners, but a significant number of them have fallen into oblivion. Most of the celebrations had a religious slant or were filled with religious elements, even for birthdays.

This religious foundation facilitated their formalisation in national organisations. The main impetus toward formal organisation came from outside, however. Between 1910 and 1920, the Arya Samaj, a fundamentalist and nationalist Hindu movement, was active worldwide. Its missionaries also visited the Caribbean (Samaroo 2006). Followers of the Arya Samaj were active in Suriname from 1912 onwards and caused friction with extant Hindu currents because of their activist stance. As a result, followers of the various Hindu denominations organised themselves into the Sanatan Dharm in 1929. The Aryans organised nationally a year later. The Hindostani Muslims, the bulk of whom were Sunnis, also organised themselves around the same period (De Klerk 1953, p. 193ff).

Heritage making included storytelling such as fairy tales, parables, religious stories and stories about the reasons for not returning to India or regret and sorrow about staying in Suriname. This sadness can still be felt by the third and fourth generation of Hindus who knew their grandparents personally. It is a broad transmission of culture via the oral route. It also included the transmission of family games (Bhagwanbali 2010; Van Kempen 2002). In this, nostalgia, suffering and grief were transmitted to the next generation in addition to family culture.

British India was thus naturally present in the lived culture of Surinamese Hindostanis. This was reflected in the crops grown, food, language, festivals, rituals, music, dances, religions and stories. To this list could also be added clothing and jewellery, but these gradually lost their specificity (De Klerk 1953, pp. 211–213). The erosion of the caste system, which was no longer a structural feature of the Hindostani community, was also conducive to ethnic group formation.

The fact that the cultural elements retained in Suriname were not a complete copy of the original culture, but were often adapted, does not detract from British India as a source and frame of reference. The creation of ethnicity, such as through song lyrics, also referred to British India and reinforced the group's connection to its motherland. From the outset, the ethnic group formation of the Hindostani community was partly determined by this connection to British India (Gowricharn 2013). In this context, Fokken (2018, p. 226ff) notes that homemaking was also promoted by housing, land ownership, mosques and temples, food, religious gatherings, public festivals, street names and parades.

Finally, the Hindostani community profiled itself as an ethnic community vis-à-vis the other ethnic communities after World War II, while VHP ideologues (Azimullah et al. 1963) as well as pandits, radio announcers, singers, poets and party activists were important actors in the launching of a civilisation offensive. This consisted mainly of propagating a family life characterised by the central position of the woman/mother in the family who stands for harmony, which is symbolised by the Hindu deities Sita and Laksmi and expressed, for example, in the Diwáli festival. With this image of women and the family life based on it, Hindostanis had to distinguish themselves from the lifestyles of other ethnic groups. It became, especially in the three decades after World War II, a normative hallmark of Hindostani life, especially of Hindostani women, a morality reinforced by Bollywood films (De Koning 1998).

The development of the Hindostani community was not a reaction to external threats as the theory of reactive ethnicity would have it (Cornell and Hartmann 2007). True, Hindostani leaders did reject assimilation at an early stage and urged the immigration of more women to enable family formation (De Klerk 1953), but it was a preference for one's 'own community', a striving for the development and preservation of primordial ties, rather than a reaction. This preference—or ethnic chauvinism, if you like—did not exclude internal divisions. Indeed, internal divisions have marked Hindostanis from the very beginning. Both elements, primordially ethnic ties and internal divisions, will affect their political participation.

9.3 The Background of the VHP

By far the most important Hindostani political party is the VHP, or the United Hindostani Party, which was founded in January 1949. Two aspects should be highlighted: its origins and its context. As regards its origins, judging from the names of the leaders of the community, the VHP emerged from an embryonic elite that was entrenched in Hindostani associational life (De Klerk 1953; Mitrasing 1959; Sedney 2017). This elite had already experienced friction with the incumbent light-skinned Creole administrators for several decades, due to the blockade to recognising the ritualistic marriage of Hindostanis and the attempts to exclude them from participating in the colonial parliament. The naming of associations like the British Indian Immigrants Association (1910), Nawa Yuga Oeday (1924, New Era) and Djagaran (1943, Vigilance) revealed their feelings. The cultural and political orientation of this elite was expressed quite early in the settlement history of the British Indians.

The context of political party formation was shaped by the Second World War. Indonesian nationalists found an ally in the Japanese occupiers in their struggle against the Dutch coloniser. This development led to Queen Wilhelmina's famous radio message in December 1942 in which she promised the colonies constitutional reforms, including greater autonomy in domestic affairs. According to several authors (Dew 1978; Ooft 1972), the promise was intended to meet the Indonesian nationalists and discourage cooperation with the Japanese. But Mitrasing pointed out that the promise stemmed directly from the Atlantic Charter, a joint declaration by the US and Britain launched in August 1941, in which these countries stated that '...they respect the right of all peoples to choose the form of government under which they will live; and they wish to see sovereign rights and self-government restored to those who have been forcibly deprived of them' (United Nations Textbook 1954, p. 21, quoted in Mitrasing 1959, p. 10).

However, while the message was addressed to the Indonesians, the commitment of the Dutch Queen matched the aspirations of the Surinamese political elite. In response to the promise, the administrative elite in Suriname organised themselves into the Union Suriname, which strove for domestic autonomy with the motto *Baas in Eigen Huis* (Home Rule). This was partly a reaction to the disturbed relationship with Governor Kielstra, and partly because of his demand that district commissioners should have some years of experience in the Dutch East Indies and (against the wishes of the Creole elite) that some Hindostanis and Javanese should be appointed as members of the Koloniale Staten (the colonial parliament) (Ramsoedh 1990). The establishment of political parties was a preparation for the promised independence (Dew 1978, p. 52; Van de Walle 1975, pp. 120–139). Anticipating the first general elections (1949), Hindostanis and Javanese set up the Hindostani-Javanese Central Council in 1946. This council later fell apart and the Hindostanis continued as the VHP.

An electoral system was devised in which the (Creole) city of Paramaribo represented a proportionally larger number of seats than the districts. The possibility of gaining some administrative control through universal suffrage produced a motley collection of participants (see Annex 9.1). Practically all major political parties were ethnically or religiously based. Also notable was the participation of some specifically agrarian interest groups, but these did not survive the subsequent elections (see Sedney 2017, p. 194ff).

The ethnic basis of the VHP can easily be explained as a reaction to frictions with light-skinned Creole administrators, i.e. reactive ethnicity (Dew 1978; Sedney 2017). The implication of this thesis is that if there had been no 'antagonism' between ethnic groups, the Hindostani community would not have produced a VHP. Thus, from this perspective, the main forging force lies outside the Hindostani community. An alternative explanation for the political unity of the Hindostani community has been proclaimed by, for example, Hoppe (1976) and Ramsoedh (2008), who believe that pattern-client relationships were the binding forces of the VHP. According to this explanation, voters receive jobs, scholarships, houses and similar favours from their political patrons in exchange for their vote during elections. Thus, the political unity would not represent a 'natural unity' but would be rather contrived and enforced by dependency.

Both explanations are debatable. The theory of reactive ethnicity ignores the primordial nature of the internal ties within the Hindostani community that had emerged long before the formation of political parties. Indications that there were widespread frictions with (black) Creoles at the time are non-existent (Gowricharn 2013). True, there were frictions with the light-skinned elite that dominated the country's administration, but these may have only reinforced the extant political cohesion of Hindostanis. Thus, very diverse groups of Hindus—Muslims and Hindostanis of different currents, of different castes, of different ideological orientations, families with different economic interests—were nevertheless bundled into one political party. The naming at the time reflected this bundling: the United Hindostani Party (see also Mitrasing 1959, p. 85).

The leadership of such a party cannot have been anything but a compromise among all the subgroups with sometimes conflicting interests (Photo 9.1).

This leadership, with the exception of its religious pillars, was not agrarian in orientation but mainly urban, with a strong leaning towards ethnic unity, entrepreneurship and education. During the 2010–2015 period, discussions were held with

Photo 9.1 Jagernath Lachmon, leader of the VHP from 1949 to 2001 and exponent of the Hindostani urban elite

his biographer Evert Gonesh, Lachmon's close party colleague Nannan Panday and fellow politician Jules Sedney, among others. From these, the picture emerged that the VHP leader, Jagernath Lachmon, was highly urbanised. He had been educated by the light-skinned notary De Miranda, was known for his preference for women of ethnic mixed descent, liked to spend his free time in the company of the light-skinned elite, who at the time met mainly in the society club Het Park, and communicated more with his colleagues from other ethnic parties than with the Hindostani constituency (Gonesh 2015). This urbanisation did not apply to the entire party leadership, but Lachmon's example was copied by other urban-oriented Hindostani politicians in the party. This is by no means to say that the VHP had no supporters or representatives from the agrarian sector, but they constituted a minority that did not outweigh the urban interest groups.

Van Dusseldorp (1963b, p. 52) calculated, on the basis of the Second General Census, that in 1950 about 16% of Hindostanis lived in Paramaribo. Vroon (1963, p. 30) worked out that around 1950, 60% of the entire population lived in rural areas. Of this, 71% lived within a 40 km radius of the capital, bringing the total population in and around Paramaribo to 83%. Moreover, the increasing urbanisation of Hindostanis is highlighted early on in history. For instance, McNeill and Lal (1915), Karsten (1930) and De Klerk (1941) point to the growing entrepreneurship and the increasing number of homeowners and the rise in the educational level of Hindostanis in and around Paramaribo (see also Fokken 2018). Urbanisation and economic differentiation formed the basis of the Hindostani elite entrenched in the VHP.

In the 1950s and 1960s, VHP ideologues developed concepts such as 'unity in diversity' and 'preservation of the cultural heritage' (Azimullah et al. 1963) underscoring not only ethnic group unity but also equality with the Creole ethnic group. Lachmon himself had 'fraternisation' as his own political concept and was focused on avoiding open confrontations with the Creole population group, always compromising with political allies and always accepting a subordinate position, although ballot box results would justify heavier claims. Despite his ideology of fraternisation and his urban orientation, he was ambivalent towards Creoles. Former Prime Minister Jules Sedney wrote in the *Suriname Weekly* newspaper (24–30 June 1989, p. 7): 'Lachmon is exceptionally tolerant towards other ethnic groups. Especially towards Creoles. Even if they are against him, it doesn't bother him. The Hindostani, on the other hand, who does not want to follow, but wishes to go his own way, need not count on understanding.'

The fraternisation strategy led to three splits until independence in 1975, only one of which had an agrarian base. The Nickerie Independent Party (NOP) originated in the mid-1950s and aimed to promote the interests of the district of Nickerie, which had become the rice barn of Suriname. Since that district was predominantly a rice district, it is safe to assume that the NOP was working hard for the rice farmers. It remained a small and not very influential political party. The Action Group came into existence in 1961 because, in short, it felt the VHP was too lenient towards the Creole group. The Action Group disagreed that mainly Creole featured in national symbols, especially the first verse of the national anthem and the kotomisie (a Creole dress) on the banknote. In addition, it had sharp critics on internal democracy of the VHP. The third split was the Hindostani Progressive Party (HPP), which emerged in the early 1970s. This group also consisted of disgruntled people who felt that the VHP was inadequately guarding its Hindostani character and stifling internal party democracy. Both the Action Group and the HPP briefly participated in coalitions as minority parties, but they did not have long political lives (Dew 1978; Sedney 2017).

It is important to underline that not only the VHP but also the Action Group and the HPP were strongly urban oriented (Dew 1978). Van Lier (1977, p. 296) notes that the political parties in Suriname had no ideological orientation and did not have a party programme either. But the practical representation of interests could force an orientation towards a particular sector. The VHP could be expected to put agricultural interests first because its electorate consisted largely of farmers. And although an agrarian focus and an urban orientation need not be mutually exclusive, such a combination of interests did not characterise the VHP. The party was carried by an agrarian constituency but led by an urban elite.

9.4 Reforms and Planning Policy

In 1943, Governor Kielstra installed a Commission to Study the Postwar Problems, which issued a report 2 years later (Rapport 1945). The commission believed that further development of agriculture and cattle breeding should be sought in conjunction with future industrialisation (Rapport 1945, p. 17). Without advocating an autarkic policy, it recommended the expansion of domestic food crop production and industrial development that should be based primarily on locally obtained raw materials, especially from the agricultural sector. Industrialisation should also be geared towards the substitution of imports (1945 Report, pp. 18–23). This was the first report in Suriname's economic history to advocate the development of an industry other than agriculture or mining.

The committee also pointed out some problems concerning industrial development in Suriname: wages were too high, labour productivity too low, the labour force too small, the availability of qualified personnel insufficient and the domestic market too small. The much-heard objection that Suriname was unfavourably situated on international transport routes was circumvented by the commission by focusing on industrialisation using local raw materials (1945 Report, p. 17). Remarkably, the commission hardly addressed the consequences of increased imports and a negative trade balance in the event of any industrialisation as Latin American countries had already experienced them by then (Klein 1946, pp. 4–5). Immigration of labour was, for the Suriname commission, a necessary condition for industrialisation. In the 1950s, attempts were made to initiate such a flow of workers, but to little avail.

In 1947, the Welfare Fund was set up to prepare for a large-scale approach to Suriname's problems. The fund's financial resources were mainly used to finance a number of agricultural experiences, improve infrastructure and inventory the country's economic aid resources. The latter included a population census, a (failed) agricultural census and the exploration of natural resources. In 1951, a Planning Bureau was established to prepare economic policies to be implemented. In 1957, the Central Bank of Suriname was established. It took over the tasks of a circulation bank from De Surinaamsche Bank, which had previously fulfilled the role of central bank.

Development cooperation with the Netherlands was among the constructions to finance development plans. An agreement was reached that one-third of the loans were provided as grants. Development aid imports became the second most important source of foreign exchange after bauxite exports. In the following period, three development plans were implemented: the Ten-Year Plan (1955–1965), the First Five-Year Plan (1965–1970) and the Second Five-Year Plan (1970–1975). In these plans industrial development was less and less in the foreground.

Economic development was discussed in detail in two publications issued by the Planning Bureau and a World Bank report (Grondslagen 1952; Surinam 1952). In the report by the Planning Bureau, the emphasis was on the agricultural sector. Although it did not reject industrial development, the narrow margins for it were outlined in detail. It also pointed to the positive effects of import substitution and industries that processed local raw materials. To encourage companies, it suggested introducing 'tax holidays' (packages of tax exemptions to entice companies to invest) and exempting them from import and export duties. Government participation in private companies should only be allowed if it reduced the capital risk for a foreign company, and if over time the shares were divested to the private sector (Grondslagen 1952, pp. 103–107).

The World Bank report mainly proposed an investment programme for the next 10 years. The emphasis was on public investment, especially for public utilities. Little was said about specific industrial promotion (Surinam 1952).

A few years later, a new memorandum was issued by the Suriname Department of Economic Affairs that advocated industrialisation (Ontwikkeling 1956). The reason was that the population growth was not being absorbed sufficiently. Registered unemployment amounted to nearly 4000 people out of an estimated labour force of 71,500. The calculation did not take into account seasonal and hidden unemployment (Ontwikkeling 1956, p. 23). The department calculated that Suriname would need 20,000 to 25,000 new labour places in the next 10 years. With regard to industrialisation, however, the report got little further than formulating some previously proposed measures to promote business establishment and the need to increase labour productivity. The department further noted that the government apparatus was inadequately equipped to pursue an effective industrialisation policy (Ontwikkeling 1956, supplement 1–3).

The preceding outline reveals that the industrialisation idea increasingly lost political weight in the decade starting from the mid-1940s. While that aspiration was initially focused on the study of the post-war problems, by the time of the Welfare Fund, the idea was hardly discussed. The plea for accelerated industrialisation in the 1950s was also not reflected in the Ten-Year Plan (TJP) that covered the period 1955–1965. Here, the emphasis was on the mechanisation of large-scale agriculture, the promotion of middle-class agriculture and the construction of a hydroelectric power station at Brokopondo for mining purposes. It was assumed that the relatively cheap production of electricity would also benefit industrialisation (Adhin 1961, pp. 145–147). However, by far the most important reason for the construction of the power plant was the need for energy for bauxite production and its processing into alumina and aluminium, an aspect that the World Bank mission had emphasised (Surinam 1952, pp. 204–215). With the realisation of this centrepiece of the Ten-Year Plan, the mining economy, rather than an agricultural economy, was consolidated.

Vroon (1963) pointed out that the TJP had a triple objective: a regular increase in the real per capita income of the population, improvement in the balance of payments and broadening of the country's economic base through which employment would also be expanded. He defined the agricultural problematic as threefold, comprising: low income relative to available land as only 60% was under cultivation, although this percentage varied greatly by region, racial group and available acreage; the one-sided planting with rice, which resulted in unemployment between labour peaks and low labour productivity; and the injudicious issuance and reclamation of the land, leading to ribbon development along the rivers and main roads, which made access to, and water control of, the entire plot impossible. Vroon (1963, p. 40) concluded that considering the land that was available, the production per hectare remained far below what was possible.

After a detailed discussion of the projects realised, Vroon (1963, p. 62ff) noted that by the end of its term, the TJP had barely met expectations. Labour supply growth was higher than the number of unemployed in the city, capital-intensive mining offered few jobs and TJP employment was mainly temporary. In animal husbandry, milk production grew but was nevertheless insufficient because milk consumption increased faster due to rising purchasing power. The same was true for meat production and the various types of fishing, such as sea fishing, inland fishing and shrimp fishing. Vroon attributed the disappointing results of smallholder agriculture to the falling prices of agricultural commodities on the world market, the heavy rainfall of the previous 5 years (he wrote in 1962), the low use of fertilisers, pesticides and seeds of adapted rice varieties, and the slow cultivation of newly reclaimed or rehabilitated land because it had not been delivered ready for construction and because the credit supply was inadequate.

The labour supply was estimated to increase annually by 2000 people (Vroon 1963, p. 70), but it is unclear whether this figure consisted entirely of workers who became unemployed or also included new entrants to the labour market. In any case, at least this number of jobseekers had to be accommodated somewhere every year. Since the private sector hardly expanded and the economy was based on a single capital-intensive mining sector, the labour supply was accommodated in the public sector on the basis of ethnic allegiances, a practice known in Suriname as 'regelen' and in the Caribbean as *apan jath* politics (Despres 1967; Dew 1978; Premdas 1996). This accommodation has characterised the political culture of all countries with a small private sector. Regelen was preceded by ethnic loyalty—it was not a bond that arose after the co-ethnics had been accommodated. To put it another way, *apan jath* politics was not an exchange relationship but a corollary of ethnic solidarity and loyalty. This practice harboured the danger that unmet expectations could turn into disappointments.

9.5 The Agricultural Programme

In post-war development plans, the VHP had the opportunity to defend the interest of the Hindostani peasants. This should be reflected in the influence of development plans in favour of the peasantry, now officially denoted as 'smallholder farmers' who formed the lion's share of the VHP electorate. Little can be said about that influence, although the VHP managed the Ministry of LVV in the 1958–1967 and 1969–1973 periods (Sedney 2017).

Mitrasing (1959, p. 86) listed the wishes of the VHP. In it, only one topic of interest appears, namely the 'promotion of agriculture'. However, it was unclear whether the VHP meant smallholder agriculture or middle-class agriculture. As for the actual influence of Hindostanis, which was most likely channelled through the VHP, I have come across only two cryptic references in official documents, both of which related to the submitted Ten-Year Plan that was published in two versions.

The first reference is Vroon's (1963), who mentioned the government's 'private discussion' of the first version of the TJP. The government critique included the fact that the Wageningen project was a Dutch concern because Suriname derived little benefit from it and should therefore be removed from the TJP. Other changes to the

TJP included the accelerated implementation of air mapping, the construction of a new airport, the establishment of a domestic airline, the construction of the East-West motorway (which, incidentally, facilitated urbanisation), the improvement of general and vocational education and of the government apparatus, and the temporary recruitment of experts in the form of technical assistance. Vroon (1963, p. 58) noted in this connection: 'These more or less incidental changes, which would sometimes have to be made under political pressure, naturally made it difficult to maintain the projected co-ordinated measures in the long term …'.

The language is very cautious, not to say woolly, even speculative, but Vroon is more specific when he states: 'Another important difference with the original plan (he was referring to the first version of the TJP, RG) was that under pressure from the Hindostani side, three times more arable land would have to be acquired, not in the districts, but in the vicinity of Paramaribo. The accompanying much larger expenditure could, it was assumed, be limited without objection by not delivering the land ready for construction and leaving the reclamation et cetera to the future landowners.' This is a remarkable passage because the Hindostani influence on the plan did not relate to the development of land in the districts or of agriculture, but to land in the Paramaribo area. This too reflects the urban orientation of the VHP, although its electorate consisted mainly of district residents. The expansion of land was probably for housing or other non-agricultural initiatives.

The second influence of Hindostanis is found in Van Lier (1977, p. 309), then director of the Suriname Planning Bureau, who was even more cryptic when he stated that 'strong insistence, especially from the Hindostani side (had been) exerted, to expand the acquisition of new land'. If this was the same reference as Vroon's, the 'strong insistence from the Hindostani side' concerned land reclamation in the Paramaribo area. Both cases involved land acquisition. If the VHP had nevertheless made an unexpected commitment to the small agriculturalists, it was to be expected that the share of the agricultural sector in GDP would have increased. This was not the case, however.

In the period 1955–1973, the VHP participated in the government twice, and in both cases it ran the Ministry of Agriculture, Livestock and Fisheries (LVV). During this period, the agrarian sector's share of GNP fell from 13 to 10%, although it increased absolutely from 16 million to 91 million Surinamese guilders. Against this, it can be argued that during this period, the share of all other sectors, except forestry and wood processing, also increased (Chin and Buddingh 1987, p. 127).

The various reports after World War II suggest that the industrialisation was to be based on a supply of labour, capital and products from the agricultural sector. This was a general idea in economics at the time (Lewis 1950). Agriculture was an important sector in Suriname, but policy was mainly focused on the Wageningen project and the promotion of middle-class agriculture. The focus was on macroeconomic development, with strong emphases on energy provision, experimental agriculture (Wageningen, Lelydorp plan and Slootwijk), export promotion and improving public finances. The emphasis on macroeconomic magnitudes left little room for smallholder agriculture, where most of the population found employment.

Vroon (1963, p. 41) pointed out the low multiplier effects of investment in agriculture because of the high import quota of national income. 'Indeed, almost all materials consumed in a polder project had to be imported: machinery, spare parts, fuels, drainage pipes, cement, etc., while wages paid were usually spent to a considerable extent on imported consumables.' The low return on investment in smallholder agriculture, the large gap between the per capita income in smallholder agriculture and the other economic sectors (Sf 350 as against Sf 3750 and Sf 8750 in industry and mining (Vroon 1963, p. 40)), the increasing urbanisation and the drive to improve the macroeconomic situation suggest that the development programme was entirely determined by the Netherlands. In particular, infrastructure and government finances were traditionally defining policy issues for the Netherlands, which wanted to get rid of the annual flow of subsidies (Gowricharn 1991). In other words, there was probably little room for political parties such as the VHP to address the importance of smallholder agriculture in larger operations such as macroeconomic planning and development.

However, it is unclear whether the VHP tried to do anything for the benefit of its agricultural constituency. Promoting the interests of small farmers, for instance, could have been reflected in a ruralisation policy of which acreage expansion would be an important indicator. Kielstra pursued a vigorous policy of ruralisation, which was part of his economic policy. The continuation of this took place in the district centre plans in which the TJP provided for land reclamation projects (Van Dusseldorp 1963a). It was during this period that the chances of achieving the ruralisation objective were greater and to which the VHP could have responded. This increased chance was mainly due to the improved road network after World War II.

On this, Van Dusseldorp writes: 'Especially in recent years, people have increasingly come to realise the importance of a good road network. This is evident when comparing the road systems of the polders, built by the government before and after the war … it appears that the accessibility of the agricultural land has improved considerably in recent years. This speaks especially when in the rainy season a clay dam is hardly passable even for pedestrians, whereas now every farm is accessible by cars practically throughout the year. This better access now makes it possible to grow crops that were hardly discussed before (e.g. plantains), as transport caused major problems in the old polders. Moreover, farmers can now move more quickly, making it less necessary to introduce new forms of habitation' (Van Dusseldorp 1963a, p. 13; see also pp. 18–19).

Economic thinking at the time was strongly marked by planning politics focused on the role of the agricultural sector in macroeconomic development (Adhin 1961; Kool 1964; Luning 1966). In that thinking, there is no complete consensus on the role of agriculture, but there is considerable overlap (cf. Gowricharn 1991; Luning 1966). Assuming a closed economy and a two-sector model, agriculture should supply surplus labour to an expanding industry, form a domestic market for the sale of industrial products and supply cheap food to industry, thereby enabling real wages to be lowered and the rate of profit to be increased, generate foreign exchange through exports and contribute to government revenues by means of taxes. At the heart of this model is the fact that the agricultural sector mechanises, releases labour

Table 9.1 Land reclamation projects in the 1950s and 1960s

Name	Year of issuance	Acreage in gross ha
Nannipolder	1954	1465
La Poule	1956	258
Weg naar zee	1956	220
Voorburg	1956	125
Welgevallen	1956	240
Tijgerkreek-Oost	1957	220
Uitkijkpolder	1957	1400
Nieuw-Meerzorg	1957	280
Mon Plaisir	1957	52
Groot Henarpolder	1958	2650
Houttuinpolder	1959	773
Leiding 6-west	1960	142
Cocospolder	–	330
Alliance	1961	300
Santopolder	1961	820
Magentapolder	1961	780
Maho	1961	280

Source: De Vries (1965, p. 128)

and lowers the price ratio between agricultural and industrial products in favour of the latter (Table 9.1).

Luning (1966) concluded that the model does not apply to Suriname. Using four criteria to assess the role of the agricultural sector, namely the production of sufficient and cheap food, the provision of labour, the provision of raw materials for industrial production and the absorption of industrial goods, and the generation of foreign exchange, he concluded that for satisfactory and accelerated economic development 'agriculture will have to make a greater contribution. This contribution will have to lie primarily in the provision of cheap, sufficient food for its own population and in the contribution through the "surplus" …' (Luning 1966, p. 134). Due to the small domestic market, only export companies were focused on 'unlimited' growth. They had to meet international quality standards, while in the domestic market the product was sold regardless of quality.

9.6 Urban Elite Versus Agrarian Electorate

When addressing the formation of the VHP, the religious and racial diversity is particularly striking. The divisions created by urbanisation and acculturation, which largely coincided with urban-rural differences and related economic contradictions, have scarcely been discussed (Dew 1978; Mitrasing 1959; Sedney 2017). Annex 10.1 shows two parties that were equally agrarian: the Agrarian Party, which consisted mainly of an ethnic mixed group of Dutch, Hindostani and Creole farmers, and the Surinaamse Landbouwers Organisatie (Surinamese Farmers' Organisation), which participated in the first elections together with a Javanese political party. The fact that two agrarian interest groups organised themselves politically indicates that they had little faith in the VHP as a guardian of agricultural interests. It is likely that the urban orientation of the Hindostani elite had already been noticed at that time. However, Hindostanis voted en masse for the VHP, motivated not by an economic interest but by the primordial ties they then possessed.

The 'neglect' of agrarian interests and the urban orientation led to a major divide that differentiated between internal ethnic Hindostani loyalties, namely between district and city. That was reflected within the Hindostani community in labels such as *boiti-* and *fótó*-coolie. Both words originate from Sranan: *boiti* stands for rural, backward; and *fótó* means city, which is higher in social status (Gowricharn 2019). The urban Hindostani in particular claimed a higher status than their co-ethnics living in the districts. The *boiti*-coolie was a peasant, in the most negative sense of the word: they were traditional, undeveloped, performed manual labour that no one wanted to do, spoke no or poor Dutch or Sranantongo, had no taste or culture and had a limited field of vision; in short, they were not urban, which for many Hindostanis was the same as 'not modern'. The term 'coolie' was already derogatory, but to be a *boiti*-coolie was the last thing you wanted to be as a Hindostani.

Conversely, rural Hindostanis replicate the fact that urban Hindostanis had been creolised: they could no longer speak Hindostani, did not know the religion and had lost their Hindostani ways of life (Gowricharn 2019). Especially after the 1950s, a silent and little-recognised cultural struggle developed in Suriname: the *boiti*-Hindostani did not want to be dismissed as a backward creature and tried to acquire normality through Western education as well as urbanised culture and modern forms of leisure such as dancing. And the *fótó*-Hindostani wanted to show in every possible way that they were modern, urban and Western or Creole and therefore culturally superior.

The urban Hindostani had existed for some time. Soon after their immigration, even before the twentieth century, some Hindostani families could be found in the city. This gradually grew to 7.7% of the city's population in 1921 (Fokken 2018, p. 263). Some occupational groups were highly mobile, for instance the milkman, the charcoal vendor and the vegetable seller. Other Hindostanis settled in the city for education or because they married an urbanite. What emerged then was an embryonic Hindostani elite that managed to work its way up in the small colonial society through entrepreneurship, education, Christianisation or by marrying members of other groups (De Klerk 1953; Fokken 2018; Gobardhan 2001).

The urbanisation of Hindostanis took two forms: Westernisation and Creolisation. The first consisted of the adoption of Dutch (and later American) culture, which took place mainly through, and was expressed in, childrearing, naming, education and the use of Dutch as a home language, in politics and urban life (for some aspects of this development, see Gobardhan 2001). The second was the result of contact with urban Creoles. During the first decades after World War II, the distinction between light- and dark-coloured Creoles was significant in cultural contact with

Hindostanis. Roughly speaking, it could be said that the light-coloured Creoles mainly contributed to Westernisation, while the black Creoles promoted blending with urban culture. In both cases, Westernisation and the adoption of urban-Creole culture, these were processes that were desired by Hindostanis; there are hardly any traces, at least in the cultural history of Hindostanis, of resistance to either form of urbanisation.

In the first three decades after World War II, urban culture was still dominated by the colonial rulers. This was reflected in the use of Dutch, its formal tone as in parliamentary speeches, newspapers and radio, and in its use as a language of social communication. This was not only due to the perceived superiority of the Dutch coloniser but also to the elite of light-skinned Creoles who, partly thanks to a formally high level of education, were strongly oriented towards the Western world. Such modernity and civilisation, in the prevailing view, could only be found in Paramaribo (Van Lier 1977). In this view, Westernisation equalled modernity and civilisation and was represented by the light-skinned Creole elite that was entrenched in civil service, political parties and churches. It broadly formed the frame of reference for the VHP elite.

With the urbanisation of Hindostanis, ethnic loyalty eroded. This was partly due to acculturation through Western education and the adoption of Creole culture. Until Suriname's independence, the political effect was that the peasants' interest was persistently neglected. In the VHP, the interests of industrialists, import traders, property owners and political entrepreneurs gained increasing weight (Hoppe 1976; Kruijer 1973; Van Waesberge 1976). As the political orientation towards the urban increased, the allegiance to the party declined further. In the 1960s and 1970s, a growing proportion of Hindostanis viewed the VHP as a party that squandered its own culture on the Creoles (even though Hindostani politicians used Hindi in their speeches to the population), had little interest in agriculture (to which the majority of Hindostanis still felt connected), was mainly a party of elite Hindostanis and was ruled by an autocratic leader, namely Jagernath Lachmon (Dew 1978; Sedney 2017).

This fragmentation of ethnic loyalties should have caused the party to disintegrate, but two developments prevented this. The first was the ethnic polarisation that began with the VHP-PNP government in the period 1969–1973. This period was characterised by racially inspired waves of strikes that strengthened the internal cohesion of Hindostanis (Dew 1978). The second development took place immediately afterwards, namely further racial polarisation in the quest for constitutional independence. This found expression in two almost entirely Creole governments, known as the Nationale Partij Kombinatie (NPK 1 and 2, National Party Combination). Both NPK governments excluded the Hindostani VHP from participating in government. In both NPK periods, there was reactive ethnicity, but it was temporary (Ramsoedh 2018).

9.7 Conclusions

This chapter has addressed the connection between ethnic and economic politics to the extent that is relevant to Hindostani farmers. Hindostani loyalties grew from micro to macro levels. This raises the question as to whether the VHP, an institution at the macro level, catered to the ethnic interests of Hindostanis, who were mainly (though not exclusively) agrarian. The party was committed to promoting unity and defending group interests. It made ideological inroads, promoted ethnic chauvinism (as did the other groups, for that matter) and, especially during elections, travelled across town and country to mobilise voters. Given this political commitment, it was to be expected that the VHP would champion agrarian interests.

The preceding analysis hardly gives reasons to confirm this expectation. First, the main intervention in post-war agriculture consisted of the establishment of the SML and the Prins Bernhard Polder. The former was initially intended to accommodate Dutch farmers and the latter had the function of facilitating middle-class farmers. Although Nickerian farmers benefited from this, these initiatives did not originate from the VHP but were proposed decades earlier by agriculturalists in Suriname (see Chap. 7). The mechanisation of rice farming was a restructuring with macroeconomic objectives (focused on exports) and produced a set of outputs of little use to small farmers. Macroeconomically, the mechanisation in Nickerie may have been a success, but macro-politically it was a disaster because it resulted in an erosion of ethnic loyalty. Such an erosion of ethnic loyalties took place not only in Nickerie but also in other agrarian parts of the Hindostani community.

For the Girmitiya population, the post-war reforms signalled two lasting developments. The first was that the agrarian human capital lost its bonding effect almost completely, either because the population deserted agriculture to settle in urban areas or because the peasants' households were transformed into farmers and became independent from kith and kin. What did persist was the ethnic community, which acted on specific social and cultural occasions as a moral community. The second development was the internal spatial and cultural differentiation between district and urban dwellers. It was a highly relevant fissure as it made it clear that agrarian interests had been neglected by the urban VHP.

It is fair to point out here that the Netherlands, motivated by purely macroeconomic objectives, had complete control over the planning. Successive governments in Suriname had been reduced to executors of the plans, probably because they lacked the expertise to offer counterproposals and because they were accustomed to following the colonial ruler. Theoretically, it cannot be ruled out that the desires of the Dutch planners and the VHP coincided. But that would mean that the VHP advocated agricultural interest. However, the idea that the VHP represented the interests of the agrarian sector is at odds with the urban orientation of the party leadership, although as mentioned earlier, there were individuals in the party leadership who represented agrarian interests. At best, the VHP was a hybrid party, characterised by an agrarian electorate and an urban leadership.

Does the VHP mean nothing to the Hindostani community? The party, like others, practised its *apan jath* politics and employed and facilitated many co-ethnics. But this policy has an intrinsic weakness. It is aimed at individual party loyalists who expect jobs, grants, licences, commissions and the like. Such an individual approach has (had) two troublesome consequences. First, it has led to arbitrariness as some families could be accommodated and others not. Partly this has to do with the social network and partly such helpful contact comes about when the 'patron' is inclined to do something. Second, an individualistic policy means that the sector in which the ethnic group is concentrated, in this case the Hindostanis in agriculture, is largely neglected. The VHP mainly offered individuals or families some social mobility (which was, by and large, true of most political parties in Suriname) and did not pursue a clear economic policy for the benefit of Hindostanis as an ethnic group or sectoral politics.

References

Adhin JH (1961) Development planning in Surinam in historical perspective. Stenfert Kroese, Leiden

Arya U (1968) Ritual songs and folksongs of the Hindus of Surinam. Doctoral dissertation, University of Utrecht

Azimullah E, Ganpat H, Lutchman W (1963) Van Brits-Indisch Emigrant tot Burger van Suriname. Wieringa, s-Gravenhage

Bhagwanbali R (2010) Hindoestaanse migranten onder het indentured labour systeem naar Suriname, 1873–1916. De nieuwe awatar van slavernij. Amrit, Den Haag

Bihari B (1974) Het jajmani-systeem in India en aspekten hiervan onder Hindustanis in Suriname. Varekamp, Paramaribo

Bürer T (1956) Het Niet–Gemechaniseerde Rijstbedrijf van 4 ha. De Surinaamse Landbouw 4(5):170–174

Chin H, Buddingh H (1987) Surinam. Politics, economics and society. Francis Printers, London/ NY

Cornell S, Hartmann D (2007) Ethnicity and race: Making identities in a changing world. Sage, London

De Klerk CJM (1941) De Brits-Indiërs in Suriname. Nieuwe West-Indische Gids 24(1):97–117

De Klerk CJM (1953) De Immigratie der Hindostanen. Urbi et Orbi, Amsterdam

De Koning A (1998) Surinaamse Sita's: veranderde invulling van vrouwelijkheid binnen de Hindoestaanse gemeenschap in Suriname vanaf 1950. EUR, faculteit Historische en Kunstwetenschappen, Rotterdam

De Vries M (1965) De Landbouwers van Suriname. Planbureau, Paramaribo

Despres L (1967) Cultural pluralism and nationalist politics in Guyana. Rand McNally, Chicago

Dew E (1978) The difficult flowering of Surinam; Ethnicity and politics in a plural society. Martinus Nijhoff, The Hague

Elahi J (2018) Baithak gáná ke nác: globalisering van een Bhojpuri volksdans. In: Baldewsingh R, Boedhoe N, Mahabier B (eds) Sarnámi Sanskirti. Enkele facetten van de Hindustaanse geschiedenis en cultuur in Suriname en Nederland. Uitgeverij Surinen, Den Haag, pp 199–210

Fokken M (2018) Beyond being koelies and kantráki. Constructing Hindostani identities in Suriname in the era of indenture 1873–1921. Doctoral dissertation, University of Groningen, Groningen

Gobardhan L (2001) Onderwijs Als Sleutel Tot Maatschappelijke Vooruitgang. Een Taal- En Onderwijsgeschiedenis Van Suriname, 1651–1975. Doctoral dissertation, University of Leiden

Gonesh E (2015) Jagernath Lachmon. Verbroederingspolitiek in Suriname. LM Publishers, Utrecht
Gowricharn RS (1991) Economische transformatie en de staat. Over agrarische modernisering en economische ontwikkeling in Suriname, 1930–1960. Doctoral dissertation, University of Utrecht
Gowricharn RS (2013) Ethnogenesis: The case of British Indians in the Caribbean. Comp Stud Soc Hist 55(2):388–418
Gowricharn RS (2019) Het spook van de koelie: de stadshindustaan. In: Baldewsingh R, Boedhoe R, Mahabier B (eds) Sarnámi Sanskirti. Enkele facetten van de Hindustaanse geschiedenis en cultuur in Suriname en Nederland, deel 2. Surinen, Den Haag, pp 107–122
Grondslagen (1952) De Grondslagen van het Tienjarenplan voor Suriname. Paramaribo, Den Haag
Hobsbawm E, Ranger T (eds). (1983/1992). The invention of tradition. Cambridge University Press, Cambridge
Hoppe R (1976) Het politiek systeem van Suriname: elite–kartel demokratie. Acta Politica 2:145–177
Karsten R (1930) De Brits–Indiërs in Suriname. Een korte schets benevens een handleiding voor de beginselen van het Hindi. Martinus Nijhoff
Klein WC (1946) Opmerkingen over Industrie in Suriname. Eswin 1:3–12
Kool R (1964) Agricultural planning in Suriname 1950–1960: An evaluation. Wageningen, Veenman & Zonen
Kruijer G (1973) Suriname, Neokolonie in Rijksverband. Boom, Amsterdam
Legêne P (1936) Het volk van het groote heimwee. Zeist, EBG
Lewis WA (1950) The industrialisation of the British West Indies. Caribbean Economic Review 2(1):1–61
Luning H (1966) Enkele kanttekeningen aangaande de taak van de landbouw bij de ekonomische ontwikkeling van Suriname. De Surinaamse Landbouw 14(4):117–135
McNeill J, Lal C (1915) Report to the government of India on the conditions of Indian immigrants in four British colonies and Surinam, Part II. London
Mitrasing F (1959) Tien jaar Suriname. Van afhankelijkheid tot gelijkgerechtigdheid 1945–1955. Doctoral dissertation, University of Leiden
Ontwikkeling (1956) De sociaal-economische omstandigheden in Suriname, in het byzonder betreffende de wenselijkheid tot versnelde ontwikkeling van industrie en nijverheid. Departement van Economische Zaken, Paramaribo
Ooft C (1972) Ontwikkeling van het Constitutionele Recht van Suriname. Doctoral dissertation, University of Leiden
Premdas R (1996) Race and ethnic relations in Burnhamite Guyana. In: Dabydeen D, Samaroo D (eds) Across the dark waters: Ethnicity and Indian identity in the Caribbean. Macmillan, London and Basingstoke, pp 39–64
Ramsoedh H (1990) Suriname 1933–1944: Koloniale politiek en beleid onder gouverneur Kielstra. Doctoral dissertation, University of Utrecht
Ramsoedh H (2018) Surinaams Onbehagen. Een sociale en politieke geschiedenis van Suriname 1865–2015. Verloren, Hilversum
Rapport (1945) Rapport van de Commissie ter bestudeering van de na–oorlogse problemen in Suriname, Paramaribo
Samaroo B (2006) The Mahatma in the Caribbean. In: Gowricharn R (ed) Caribbean transnationalism. Migration, pluralization, and social cohesion. Lexington Books, Lanham/Boulder, pp 139–153
Sedney J (2017) De Toekomst van ons Verleden. Democratie, Etniciteit en politieke machtsvorming in Suriname. Vaco, Paramaribo
Sollors W (ed) (1989) The invention of ethnicity. Oxford University Press, New York/Oxford
Surinam (1952) Surinam, Recommendations for a ten-year development program. John Hopkins Press, Baltimore
Van de Walle J (1975) Een oog boven Paramaribo: Herinneringen. Querido, Amsterdam

Van Dusseldorp DBWM (1963a) Het Centraplan voor Suriname: Some aspects of town and country planning in a developing country. Wageningen, Veenman & Zonen

Van Dusseldorp DBWM (1963b) Geografische Mobiliteit en de ontwikkeling van Suriname. Bijdragen tot de Taal-, Land-, en Volkenkunde 199(1):18–55

Van Kempen MHG (2002) Een geschiedenis van de Surinaamse literatuur. Uitgeverij Okopipi, Paramaribo, pp 279–305

Van Lier R (1977) Samenleving in een grensgebied: Een sociaal–historische studie van Suriname. Emmering, Amsterdam

Van Waesberge R (1976) Suriname, Een stukje derde wereld. Novib, Den Haag

Vroon IJ (1963) Voorgeschiedenis, opzet en resultaten van het Surinaamse tienjarenplan. Nieuwe West-Indische Gids 43(1):25–74

Yang A (1979) Peasants on the move: a study of internal migration in India. J Interdiscip Hist 10(1):37–58

Chapter 10
The Transition of Rice Peasants, 1960–1980

Abstract This chapter argues that the transformation of peasants to urbanites was a complex and comprehensive process. The agrarian problem after WW II consisted not only of insufficient income or a lack of work but included poor social and cultural livelihood opportunities for the peasants and their children in the districts. Successive development plans have hardly provided for needs such as agrarian employment, further education and decent life conditions. Alternative employment arose in the districts, making part-time farming a structural feature of agrarian existence. This argument is structured as follows. Section 10.2 discusses the demographic shifts Suriname experienced after World War II. These shifts deepened the differentiation within the Girmitiya community and rendered the residuals of the agrarian human capital useless. Section 10.3 discusses the transformation of Nickerie into the rice barn of Suriname, while Sect. 10.4 argues that the urbanisation of the Girmitiya rice peasants was also a cultural process that took place in the city and within districts. Section 10.5 discusses the emigration to the Netherlands that affected agrarian production. These developments led to a shrivelling and blurring of the Girmitiya peasantry in Suriname.

Keywords Demographic shifts · Rice politics · Demise of peasants · Emigration · The Netherlands

10.1 Introduction

Successful farmers remained in the agricultural sector, which was particularly the case in the Nickerie district. But even there, marginalisation of small farmers did occur. The main focus, however, was on the successes and returns of the large investments made there. Mechanisation in Nickerie generated two studies, one of which dealt with the Wageningen project (De Wit 1960), while the other was a study focusing on technological changes (Sital 2002). The marginalisation of smallholder agriculture has received little political or scientific attention. In the literature, it has mainly been seen as a factor that led to the departure of peasants to the city (De

R. Gowricharn, *The Girmitiya Peasants in Suriname*,
https://doi.org/10.1007/978-3-031-67961-2_10

Vries 1965; Kruijer 1951; Van Dusseldorp 1963b). But this observation has not been more than an indication of district emptying.

Like the later emigration to the Netherlands, the migration to the city from the rice districts was triggered by mobilities, along with spatial, political, economic, social and cultural changes (Schiller and Salazar 2013). Economically, poverty and the mechanisation of agrarian production were at first glance the most important causes. Both effects should provide cheap labour and capital to an industrialising city, according to the highly popular economic growth model of Lewis (1954), but in Suriname, the supply of labour was limited because part-time farming had become a structural feature of agrarian existence (Luning 1966a, p. 8). From this, one might infer that the availability of a piece of land could have made part-time farmers cheap because they provided their own food, just like the British Indian labourers who worked on the plantations at the time or were forced out as part-time farmers. In order to offer part-time labour, a farmer must be near their plot to cultivate their land and near an employer to work as a wage labourer. Geographically, this was an almost impossible combination.

But the main argument against Lewis's model is that no industrial development got off the ground in Suriname. The effect of mechanisation was that rice production increased significantly and the country met its food requirements, which Sital (2002) referred to as 'Suriname's Green Revolution'. However, the absence of significant investment from private capital and the limited supply of labour, as reflected in urbanisation, were interrupted by independence in 1975. The migration from the districts to the city was thus historically continued by a further migration to the Netherlands.

In this chapter, I argue that the transformation from farmers to urban dwellers was rather more complex and comprehensive than simply a matter of making ends meet. The agrarian problem consisted not only of insufficient income or a lack of work but of poor social and cultural livelihood opportunities for the peasants and their children in the districts. A purely economic explanation ignores the fact that several governments since Governor Kielstra had made efforts—albeit in vain—to improve the lived environment in the districts. Successive development plans hardly provided for needs such as employment, further education and decent life conditions. Moreover, the population in the residential centres remained small as children moved to the city of Paramaribo. Besides this frugal approach, alternative employment arose in the districts, making part-time farming a structural feature of agrarian existence.

In the next section, the demographic shifts Suriname experienced after World War II are discussed. These shifts deepened the differentiation within the Girmitiya community, rendered the residuals of the agrarian human capital useless but enabled the achievement of new human capital. Section 10.3 discusses the transformation of Nickerie into the rice barn of Suriname, which also brings into sharper focus the loss of the agricultural character of the other districts. Section 10.4 argues that the urbanisation of the Girmitiya rice peasants, using the concept of mobilities, was also a cultural process that took place in the city and within districts. Section 10.5 discusses the emigration to the Netherlands that affected agrarian production. These developments led to a shrivelling and blurring of Girmitiya peasantry in Suriname. Section 10.6 interprets the findings.

10.2 Demographic Shifts

Urbanisation has traditionally been seen as a problem as it has overburdened the city. This concern reflects an excessive focus on the city, as the lack of facilities in the districts and inland areas has hardly received any attention. The concern over urban overcrowding has regularly recurred in various reports since then and reached policy status with Kielstra's ruralisation policy and the later centre plans (Van Dusseldorp 1963a). But to no avail: the centres were minimal in terms of life conditions and prospects, and for further study, work or recreation, residents of the districts had to move to the city.

Due to the strong focus on the city, inter-district migration is a scarcely discussed topic. This phenomenon is also old. Already during slavery, freedmen moved from area to area (and to the city) to work. Later, free Hindostani plantation workers also moved from plantation to plantation and from district to district (Van de Houven van Oordt 1900). Still later, peasants could get a piece of land in another location and moved for that reason (Burger 1928). Some peasants exchanged residence to join the family; women moved because of marriage to join the husband's family; and both men and women could leave sparsely settled areas to settle in larger Hindostani concentration areas (Speckmann 1965). The Marowijne and Nickerie border districts experienced some migration due to employment, but this was mainly of foreign origin. Nickerie in particular had a highly mixed population before World War II, mainly from the English-speaking part of the Caribbean (Van Dusseldorp 1963b). Ignoring this immigration led to a strong undervaluation of activities in the border districts.

The transformation from farmer to urbanite with the concept of mobilities also establishes a link between those who stayed behind and those who left (Alexander et al. 2018; Schiller and Salazar 2013). Older sons and daughters often stayed behind with parents to give younger children the opportunity to work or study elsewhere. From this perspective, migration is not only a push but also a sacrifice of, or an opportunity provided by, those left behind.

Steady urbanisation was reflected in the percentage increase in the Hindostani share of the urban population. Figures on this are not detailed and do not give a complete picture. The first census (1921) reported a Hindostani share in the urban population of 7%, the second census (1950) reported a percentage of 18% of Hindostanis in Paramaribo, the third census of 1964 could not be ascertained, the fourth census (1971) does not contain data on the percentage of Hindostanis in Paramaribo, and ethnic registration was not included in the fifth census (1980) for ideological reasons (Menke and Sno 2016). Between 1972 and 1980, the population size decreased due to emigration to the Netherlands. It is difficult to say anything about urbanisation during this period. Although Paramaribo had an increasing percentage of Hindostanis, at the same time the number of Hindostanis in the districts also rose (Van Dusseldorp 1963a).

In Paramaribo, Hindostanis increasingly organised themselves into formal institutions that influenced life in the districts. That included the establishment of Hindu and Muslim schools, broadcasts through the radio stations AVROS, Apintie, Rapar and later Radika (the latter was aimed mainly at the district population and represented 'the voice of the district'), the establishment of orphanages, cemeteries and a crematorium, and an increasing number of sociocultural associations, a cinema system that mainly showed Indian films and a vibrant community life (Gowricharn 2020, Ch. III). In this process, the Indian consulate (later converted into the embassy) was of great significance with its offer of classes in Hindi, musical instruments like the dhol, tablá and sitár, classical dances, and Indian dishes like samosá, biryáni, rasgullá rasmaláy and pakorá. The VHP, which was housed in the capital and operated from there, also provided an urban frame of reference for district residents. School attendance, the purchase of scarce products (such as jewellery), the sale of vegetables, fruits and fish, visiting relatives and the like promoted urban influence. In short, although the number of Hindostanis in the city was numerically a minority, their cultural influence on the peasantry was significant.

Due to the small scale and improved infrastructure (especially after the Second World War), distances between Paramaribo and surrounding districts were easy to bridge. This made it possible for numerous urban Hindostani families in the districts to still keep a plot to plant rice and horticultural crops for their own use. Morenc (1988, p. 127) qualified these people as 'weekend farmers' who earned a full income outside agriculture and had little understanding of farming. This last addition seems to me a dissonance since many Hindostani urban dwellers were still intertwined with district life, had some access to the land and grew their own food.

Consequently, migration to the city was not always a rapid process like emigration. Rather it happened gradually. This was reflected in the proliferation of part-time positions, such as that of market woman. Often these vendors lived a long distance from the (urban) market, so they were on the road for a long time and spent entire days at the market selling their wares. Photo 10.1 shows a marketplace in Paramaribo and female vendors who must have lived in the vicinity of the city. Considering the heavy weight of the goods and the fact that some of them could spoil easily, these vendors could not afford to spend much time and energy on travelling.

Table 10.1 is a compressed version of more extensive tables in the 1959 and 1969 agricultural censuses (DLS 1969; TLS 1959). The table shows only the columns with the number of farmers (the censuses speak of farm managers) who have agriculture as their main or secondary occupation. Across the country (see last row), the number of people with agriculture as their main occupation decreased by 16% in 10 years, while the number practising agriculture as a secondary occupation increased by 20%. The largest increase in the number of people with agriculture as a secondary occupation took place in the rice districts of Nickerie, Suriname, Saramacca and Commewijne. The comparison with the two survey years reflects the continuation

Photo 10.1 Hindostani market woman in Paramaribo, 1947

Table 10.1 Main and secondary occupation of agricultural farm managers, 1959–1969

	Farm managers 1959		Farm managers 1969	
	Main occupation	Side occupation	Main occupation	Side occupation
Paramaribo	69	64	23	17
Nickerie	2107	416	2021	811
Coronie	529	157	353	161
Saramacca	1593	241	1115	432
Suriname	4879	2636	3845	3089
Para			363	363
Commewijne	1440	1618	1066	1165
Marowijne	65	71	134	178
Brokopondo			8	31
Total Suriname	10,682	5203	8925	6247

Source: TLS (1959, p. 43), DLS (1969, p. 33)

of a trend, namely the move towards the third transformation through entrepreneurship or wage labour outside agriculture.

In the tables of the agricultural censuses (1959 and 1969), the columns are broken down by farm size class. In all size classes except the largest, the number of farmers with farming as their main occupation decreased, mostly in the smaller farms (< 0.48 ha), and the number of farmers with a secondary occupation increased. The most plausible interpretation of these figures is that farm managers on the smaller plots have left farming and taken jobs elsewhere. The category with a secondary occupation is still high in the smallest class (41% of the total number of farmers in 1969), but these are probably district residents who grow rice (and other crops) for their own consumption. The inclusion of these 'farmers' in the census of agricultural holdings therefore gives a distorted picture of the number of farmers and workers in the agricultural sector.

What did these shifts mean for the welfare position of small farmers? In the mid-1960s, i.e. a decade after Bürer's studies, Luning evaluated several surveys conducted by CELOS, the Bureau of Rural Development and the Ministry of LVV, and concluded: '... one can say without reservation that on at least 80% of Suriname's farms a minimum subsistence existence is not achieved under present conditions' (Luning 1966a, p. 3). Based on his own research among farmers in the Leiding area, Nickerie and Calcutta (in the Saramacca district), he calculated that net incomes—obtained from vegetables and annual and perennial crops and income from outdoor work—over 6 months in 1965 amounted to Sf 448.50, Sf 344 and Sf 183, respectively. Note that this figure includes non-farm income. Meanwhile, the standard for a decent living had risen from 1000 to 2000 Surinamese guilders. In other words, the gap between minimum subsistence and what the farmer earned in income, i.e. the poverty level, had widened.

10.3 Nickerie: The Rice Barn of Suriname

In the TJP (1954, p. 30), Sf 29.3 million of the estimated Sf 117 million was spent on agriculture, livestock and fisheries. Of this, the largest expenditure, Sf 13.5 million, went on opening up new acreage, and the rest on improving existing acreage (Sf 4 million), agricultural credit (Sf 3.5 million), drying plants and storage places (Sf 2 million), agricultural information and education (Sf 1 million), agricultural research (Sf 1.2 million), the Bureau of National Development (Sf 2 million), fish farming (Sf 1.2 million) and cattle breeding (Sf 0.9 million). The agrarian objectives of the TJP were to produce enough food for the population and for export, and to strive for a reasonable income for the farmer and his family. Kool (1964, p. 103) gave a scathing verdict on these plans: 'Various estimates made were incorrect, as no data were available, at the time the plan was formulated. Population growth was underestimated, as were the costs per hectare of new land ... It was a kind of primitive form of an administrative budget.'

At the heart of rice policy was the Wageningen project in the Nickerie district, then one of the most advanced rice farms in the world. The set-up of this farm, formally organised as the SML, initially envisaged an exploitation of 15,000 ha of land but was limited to 6000 ha for cost reasons (De Wit 1960) and expanded to 10,000 ha by 1980 (Sital 2002, p. 54). The Wageningen project was set up as a rice export company that enjoyed all kinds of competitive advantages in the European market. Part of the production was sold on the local market.

The SML radically changed the rice production process. Typical features included the use of aeroplanes, new rice varieties, new methods of sowing, planting, harvesting and threshing, the aim of two harvests a year, and the increasing use of chemicals to control weeds and insects and the fertilising of the soil. Former Nickerie district commissioner Ramadhin (2014, p. 47) points out the downside of this process: 'A big loser of the Green Revolution in Nickerie turned out to be the environment. The use of chemical pesticides, such as insecticides, herbicides and pesticides, and the abundant use of fertilisers, led to ecological imbalances. The pathogens of diseases and pests continued to increase in size as their natural enemies were eliminated by spraying. This development caused the swamp fishery to decline sharply. The swamp fish die as a result of the use of these chemicals.' The polluted soil also made it difficult to alternate rice farming with horticultural crops (Photo 10.2).

The use of pesticides, insecticides and fertilisers, all part of mechanisation, had two main effects on farming. On the one hand, it enabled greater production while saving labour, so production per unit of labour increased. On the other hand, it polluted the soil. While in the earlier (pre-war) forms of production, the period between sowing and harvesting was still used to grow horticultural crops, on mechanised farms this was virtually impossible due to soil pollution (Sital 2002). The exception were the large farms that could leave pieces of land fallow for natural regeneration, but the majority of farmers did not consider this option. As Ramadhin (2014) pointed out above, land degradation also affected water quality and fish stocks.

Photo 10.2 The rice store in Wageningen, the centre of mechanical rice cultivation

Farmers were thus forced to strive for a higher cash income because they could no longer grow numerous horticultural products they had previously grown or keep small livestock they had kept, at least not on the same scale.

Kool (1964, pp. 109–110), who evaluated agricultural planning in the 1950s–1960s, stated: 'Rice production for local consumption rose during those years by a maximum of 20%, while population increased over the same period by at least 30%. If we exclude the output under the Wageningen scheme, we find that the results of the ten years of agricultural planning are decidedly unfavourable ... Exports by the agricultural sector would be almost doubled in the six years between 1959 and 1965, while imports by that sector would dwindle. These were the objectives of the Plan but they were never achieved.' He continued by commenting that since 1963, government policy had shifted from small-scale farms to medium-sized and large farms. But even this shift was based on conjecture because it was unknown whether the farms were technically and economically viable.

Criticism also came from the political left. Students from the Netherlands, especially from the agricultural University of Wageningen (the Netherlands), pointed out that the research of the Prins Bernhard polder (200 ha in size) remained in Dutch hands. The consequence of this, as Hasselbach (1966) argued, was that the research took place mainly on large-scale farms and that the results were difficult to translate to small farmers. The Wageningen project, which accounted for 36% of national rice production in 1968, also negatively affected rice farmers in the Nickerie district by exporting 84% of its production and selling the rest on the local market. Moreover, the SML acted as a buyer of the crop, supplying seeds, fertilisers, water and the like at a certain price. This yielded a profit of three-quarters of a million Sf for the year 1968/1969, for example (Uitbuiting of Suriname 1970, p. 77). Unlike the rice-exporting farmers, the SML obtained export licences easily and could take advantage of the discounts on the EEC market. Six of the company's nine supervisory board members were appointed by the Dutch government, the remaining three by the Surinamese.

Nevertheless, the SML and the Prins Bernhardpolder project had beneficial effects on local farmers by providing their expertise and facilities (Ramadhin 2014, p. 48). The SML also collected a large amount of experience and data for the purpose of reclamation and extraction of virgin land (Sital 2002, p. 55). Applied rice research conducted in these large-scale farms, according to Sital, focused on '...testing the suitability of different types of imported farm machinery and equipment for (and adaptation to) the local wetland conditions and the requirements of the rice crop; land development and water control involving field layout, land levelling, irrigation and drainage facilities; and development of new rice producing technologies involving new varieties, new (mechanized) cultural practices and the processing of rice'. However, it was mainly larger farmers who sought, tested, selected and adapted new practices and rice varieties and incorporated them into their existing operations. The spread of new techniques took place mainly through the experience gained in networks of families and friends. In fact, there was a period of 10–15 years between the availability of these new techniques and their adoption (Sital 2002, p. 56).

Based on a sample of farms, Sital (2002) described the process of mechanisation in the period 1967–1970 in the Nickerie district. That process consisted of an evolution from manual tillage followed by draft animals and tractors; the adaptation of rice varieties that resulted in higher yields per unit of land that could be harvested using combines; rice varieties that were qualitatively superior to the older varieties; the use of fertilisers, the adoption of new cultivation techniques, better water control, machine harvesting and threshing and more intensive land cultivation. Although mechanisation in Nickerie was more advanced than rice cultivation in the other districts, Sital (2002, p. 57) points out that many farmers (in the Groot Henarpolder it was 63%) earned an income off-farm. Most of these farmers found employment in the Prince Bernhard polder and the Wageningen project. 'By the late 1960s a twofold job was almost a common practice among rice producing farm families in the district' (Sital 2002, p. 57).

Although all rice farms in Nickerie were improving, the need to earn supplementary income was compelling. Sital created a typology of rice farms by size of plots: 1–2 ha, 4–6 ha, 8–12 ha and equal to or larger than 12 ha. In all size classes, families were forced to earn income from wage labour. At the end of the 1960s, rice production on 1–2 ha farms contributed 20% of the net family income, while on 4–6 ha farms it was 39%, on 8–12 ha farms it was 58% and on farms equal to or larger than 12 ha the 'rice income' was 72%. The rest was earned from growing other crops, other activities or from wage labour (Sital 2002, p. 252). Looking back at technological innovations and their effect on household incomes, Sital (2002, p. 254) noted: 'The actual technological changes and improvements in rice farming up to the early 1980s did not affect the (real) farm business income earned with modern rice farming in the late 1960s, nor did they widen or narrow the gap in this income between the larger and the smaller farmers.' In short, Suriname's 'green revolution' was limited.

Production in the old rice districts declined sharply in the first half of the 1970s but was offset by sharp increases in Nickerie. Ramadhin (2014) attributes the decline to the NPK-1 government (1973–1977). The then minister of economic affairs, Eddy Bruma, wanted to provide cheap food for the urban population. 'In that period, every rice exporter had to apply for a permit from the Ministry of Economic Affairs to be allowed to sell rice export ... The minister ... required the exporter to sell some of the rice below cost price to the government ... Quite simply, the rice farmers had to subsidise the urban population with cheap rice' (Ramadhin 2014, p. 47). A similar observation is made by Rest (1974, pp. 15–16), who writes that the neglect of smallholder agriculture is due to Creole political parties that had no interest in fostering the economic power of Hindostanis. In the government policies implemented, priority was given to the urban Creole electorate.

From the mid-1950s at the start of the TJP to the 1980s, paddy production in Nickerie increased dramatically, as Table 10.2 shows.

By the mid-1970s, almost all paddy production (95%) came from Nickerie. However, a significant proportion of this was accounted for by large-scale agriculture, including the SML, the Prins Bernhardpolder and Van Dijk's farm. Large-scale agriculture in Nickerie accounted for 61% of the total district production and 56%

Table 10.2 Nickerie's share of Suriname's paddy production, 1954–1981, in tonnes

Period	Small-scale agriculture	Large-scale agriculture	Total Nickerie	Nickerie as % of Suriname	Suriname total
1954–1956	21,065	7190	28,255	42	67,473
1959–1961	22,510	22,772	45,282	59	77,072
1964–1966	31,212	31,922	63,134	69	92,016
1969–1971	47,723	59,481	107,204	81	131,672
1974–1976	60,055	101,211	161,266	95	169,921
1979–1981	91,375	144,793	236,168	92	258,043

Source: Compiled from Sital (2002, p. 39)

of the production of all districts. However, smallholder agriculture (read: rice farming) in Nickerie, which was highly advanced compared to the other districts, was not a uniform culture. It was divided into different classes of farms that differed by plot size, the availability of family labour and the degree of adoption of new technologies (Sital 2002). Larger farmers were better educated and had larger families with more adult members. But these larger farmers were not always better off than the smaller ones. Sital commented (2002, p. 254): 'The smaller farms were operated predominantly by nuclear families, the larger farms increasingly by joint families. This means that the incomes on the larger farms were not only earned but also had to be shared by, on average, more nuclear family units than on the smaller farms.'

Sital explained the exceptional position of Nickerie in terms of historical and geographical circumstances. The district is located in the far west of Suriname. It has long been geographically isolated from the rest of Suriname, separated from neighbouring Guyana only by the Corantijn River. As a result, Guyanese workers came to work in Nickerie, marriages between Nickerians and Guyanese took place, and sales markets of rice and other products in Guyana were closer than Paramaribo (see also Singelenberg 1983). In short, for Nickerians, Guyana was geographically, economically and culturally closer than the rest of Suriname. It can be added that water control was easier in Nickerie, while land was more fertile and allowed for easier expansion than in other districts (Luning 1966b, p. 130).

After independence in 1975, the Wageningen project was handed over to the Surinamese government. According to Ramadhin (2014), the project has since gone backwards, mainly due to the appointment of poorly skilled managers. The company slowly built up unaffordable debts and eventually went bankrupt, but it is unclear whether mismanagement is the sole cause of this company's demise as many Nickerians tend to think. Market conditions may also be among the causes of the bankruptcy. Since its establishment, the company, like all rice exporters, has benefited from preferential access to the European market. Added to the relatively high labour costs in the agricultural sector in Suriname, it is very likely that despite mechanisation, rice exports were not competitive on the international market.

10.4 The Differentiation of Peasants

To put developments in Nickerie in perspective, it is important to note a differentiation of rice farms. In previous studies, Luning (1967, 1968), Luning and Sital (1979) and Morenc (1988) pointed to two differentiating factors. The first is that rice culture had a wide variety that allowed classification into traditional, transitional and modern rice farms. This typology, incidentally, varied with the ethnicity of the farmer, the size of the plot, the crops grown and the location in relation to marketing markets. Second, the 'rice income' was insufficient to make a living. This was a tradition that was already known from the colonisation of British Indian labourers who had to work part-time on nearby plantations.

The duality of income from one's own plot and externally earned income had become a permanent feature of peasant life, an agonising concern that authors in the early twentieth century had also pointed out (Leys 1921). During and after WWII, studies made it clear that most peasants could not make ends meet with the income earned from agriculture. And in the 1960s and 1970s, it appeared that even the larger farmers in Nickerie relied on income earned outside the rice sector. The conclusion inevitably is that the size of the cultivated plot, although positively related to family income, does not sufficiently explain the undignified conditions in which the bulk of rice farmers have always found themselves.

A second conclusion flows from the Nickerie data. If 90–95% of paddy production was produced in that district, the rest came from other districts. This inference touches on the implicit assumption that every Hindostani district resident who had access to land was a paddy farmer. Peasants outside Nickerie (together accounting for 5–10% of paddy production) produced mainly for their own consumption. The association between agriculture, horticulture and living in the district was broken. Migration to the city was thus one form of the transformation from peasant to urban dweller.

Another form of transformation took place within the districts themselves. It is important to dwell longer on this 'internal transformation' of the districts. This is best done by considering the terms 'part-time' and 'full-time' farmers against each other. Morenc (1988, p. 125ff), following the official definition, interprets a full-time farmer as one working in agriculture at least 3 days a week. In 1960, only a third of the population was working as farmers (in arable farming, animal husbandry, horticulture, fishing, hunting and/or forestry).

By 1976, that percentage had declined to 16.5% and it further declined to 9.18% in 1981, which, according to Morenc, should be adjusted to 6.8% based on LVV surveys. This decline in the agricultural population was accompanied by a sharp increase in the number of part-timers, i.e. people who worked less than 3 days in the agricultural sector. The drop from one-third of the agricultural population in 1960 to 6.8% in 1981 reflects a spectacular decline. This can perhaps be partly explained by emigration to the Netherlands, but assuming that emigrants left from all sectors, the percentage difference between sectors must not have been very large.

Also interesting is the observation by an LVV researcher that socially weak people unable to cultivate their plots or find work outside agriculture count as full-timers, although they work fewer hours than part-timers (Heybroek 1978, p. 26). Morenc (1988, p. 127) adds: 'If they find a job outside farming, they keep the plot or rent it out. With farming they usually stop or it had to be very extensive crops, such as cattle breeding and tree cultivation.' The term 'full-time' is thus contaminated and includes fewer farmers than mentioned in the agricultural censuses. In the 1970s, the share of full-time farmers in agriculture was less than 6.8%.

In the late 1970s, 90% of Hindostani farmers with plots of 1–2 ha had a secondary occupation, while among Javanese farmers it was 100%. On farms of 4–6 ha, part-time employment was 55%, and on those between 8 and 20 ha it was 66%. 'In the districts, one finds employment in the ministries of Home Affairs, District Administration, Public Works and Transport and Agriculture, Livestock and Fisheries. In 1980, for example, of Saramacca's economically active population, 47% of men and an equal proportion of women worked in local governments' (Morenc 1988, p. 129). Morenc notes that a civil service job offered the opportunity to do other work 'in the boss's time', for example as a transporter of people and agricultural products, tailor, gold- or silversmith or shopkeeper. To this one could add other professions such as carpenter, mechanic and the like. These activities, like the civil service jobs, did not de facto have to be full-time jobs. Civil service jobs were often little more than formal appointments, an accommodation. The transformation was thus a variant of 'flight from agriculture' without moving to the city.

Due to the aggregation of very different categories of farmers as part- and full-timers, it is impossible to get a good picture of their age structure. The farmer registered as full-time could be the driver of a scheduled bus, and the part-timer could be a full-time civil servant or shopkeeper. The second and third agricultural censuses (see Annex 10.1) show statistically few differences in the age structure of farm managers: in both years, the proportion of over-55s was about 31%, while the youngest age category (< 35 years) saw a drop from 26% to 20%. Business succession had already been highlighted by Beck (1924) as a major continuity problem

The observation of an aging and poverty-stricken group of rice farmers is also regularly raised in later literature (Morenc 1988; Sital 2002; Van Dusseldorp 1963b). Many young people left agriculture for further education or to get jobs elsewhere. This was also true for those who had received agricultural education, which mainly attracted urban children who found employment in the government, trade and large agricultural companies. In other words, the transfer of agricultural knowledge did not find its way into the production process. Like many training programmes in Suriname, agricultural training had the function of giving the diploma holder access to a white-collar job that offered more prestige and a more secure income (Bovenkerk 1975; Gobardhan 2001; Van Lier 1977).

Although business training and succession was a concern, rice production in Suriname was often approached as an economic issue, and more specifically as a market issue. In this perspective, quantities, productivity, quality, prices and marketing were central. In addition, insufficient land (both quantity and quality), inadequate agricultural knowledge, poor water control and the role of buyers would explain the poor position of farmers. The solution was sought in cooperatives of all kinds that had to express the agency of the peasants. But this role of cooperatives was disappointing, as Table 10.3 shows.

In nearly 70 years, 90 cooperatives were established, of which only 26 were active at the time of the survey (Morenc 1988, p. 293). This does not exactly indicate success. The approach in terms of market regulation, especially the setting of minimum and maximum prices, is also insufficient to make agriculture attractive. Even with extensive mechanisation, as in Nickerie, there was still a need to earn supplementary income outside rice farming.

This Janus face of mechanised rice farming had led Luning (1966a) to propose that the duality of peasant existence should not be fought but rather taken as a starting point of a structural situation. What such a policy should look like remains an open question. In any case, additional employment will have to be created in the immediate vicinity. This is only possible if public services are opened, private entrepreneurs create jobs or peasants themselves develop new activities.

Not only did rice farming transform, but people changed too. Speckmann (1965, p. 243) reported that as a result of the modernisation of rice farming, traditional mutual unpaid help by neighbours and relatives had declined. Kruijer (1949) had pointed this out earlier. A similar observation was made later by Morenc (1988). In addition, the decrease in time investment had two other effects: women were less and less present in the production process (rice farming became a male affair) and children were given more space to attend school. Although Morenc only touched on the effects of urbanisation, the migration to the city created a labour shortage among a growing category of socially needy people in the districts, such as widows and elderly farmers, marginal farmers and part-timers with fully fledged secondary occupations. That labour shortage ranged from 2.5 to 3 h a day, i.e. a quarter to a third of working time (Morenc 1988, pp. 121–22). Only horticultural farms could still run entirely on family labour.

Table 10.3 Cooperatives formed between 1914 and 1983 in the agriculture, livestock and fisheries sectors

District	Agriculture	Animal husbandry	Fishery	Combination	Total	Non-functional
Paramaribo	4	2	1	–	7	6
Suriname	16	–	1	1	18	11
Para	1	–	–	–	1	–
Saramacca	10	–	1	3	14	11
Brokopondo	1	1	–	–	2	–
Coronie	3	2	–	–	5	3
Nickerie	27	–	2	5	34	26
Marowijne	1	–	–	–	1	–
Commewijne	6	–	1	1	8	7
Total	69	5	6	10	90	64

Source: Morenc (1988, p. 293)

In the decades after World War II, the lived environment in the districts was urbanised. As part of the development plans, roads were built and sometimes paved, water supply and electricity became available in an expanding circle around Paramaribo and the district capitals, while the number of schools, medical and police stations, and government offices outside the city also increased. Photo 10.3 shows the first signs of district urbanisation. Not only do the roadside ditches and the wooden bridge of neatly tailored poles and planks look modern, but the electric wooden poles and wires as well as the houses with zinc roofs stand out. They reflect urbanisation and prosperity.

In the 1960s and beyond, Hindostani radio, and later television, appeared (Gowricharn 2020, Ch. III). The number of Hindostani cinemas in the districts increased. With this urbanisation of rural areas, the reported numbers of people employed in rice farming say something, to a decreasing extent, about the agricultural base of the Hindostani community. Living in the district did not automatically mean being a farmer. The declining interest in agriculture was also facilitated by the education that prepared the district resident for a living in the city.

Photo 10.3 Beginning urbanisation of the districts

10.5 Emigration

Shortly after the NPK government took office in 1973, it announced its intention to achieve independence 2 years later (Ramsoedh 2018). In retrospect, this aspiration could have been foreseen, but the announcement shocked the country as it was not an election issue. Hindostanis in particular felt unsafe, not only because Creoles insisted on assimilation and saw Hindostanis as unwelcome competitors but especially because with independence, the protection of the Netherlands would disappear. The independence announced in 1973 suited the Netherlands, which had also expressed this desire a year earlier.

Meel (1994, p. 657) mentions three reasons why the Netherlands strove to grant constitutional independence to Suriname: the labour unrest in the Netherlands Antilles in 1969 had led to the deployment of marines, something the Netherlands wanted to avoid in Suriname; the growing image of the Netherlands as a colonial power because parts of the country still possessed a semi-colonial status; and the swelling flow of Surinamese immigrants to the Netherlands, which led to growing unrest domestically. As a result, in 1973 an 'international grand coalition' emerged between the Dutch and NPK governments to push through independence at all costs.

The emigrants consisted mostly, but not exclusively, of Hindostanis. While emigration was taking off, the Surinamese Ministry of Social Affairs conducted a survey (Zielhuis and Girdhari 1973). The timing was unfortunate because migration had not yet reached its peak. Moreover, the researchers limited themselves to determining country of origin, age, sex and motive for departure. After independence, Surinamese still had the opportunity to come to the Netherlands as a part of family reunification. Mügge (2012, p. 5) reported that the number of Surinamese people in the Netherlands in 2010 was 342,016, 156,947 of whom belonged to the second generation. This meant that 185,969 belonged to the first generation, i.e. people who had moved to the Netherlands from Suriname.

Choenni (2016, p. 37) arrives at a lower estimate: in the period 1973–1975, according to him, 68,700 Surinamese would have left their country, a fifth of the then Surinamese population. Of these, 40,000 would have been of Hindostani origin. In 1972, the Algemene Bureau voor de Statistiek (ABS, General Bureau for Statistics) reported a figure of 142,300 Hindostanis in Suriname for 1971. After independence, a less substantial emigration took place until 1980 in the context of family reunification. While Choenni's figures relate to a limited period (1973–1975), Mügge's figures cover the entire emigration of Surinamese, i.e. from World War II until the end of family reunification in 1980. It is clear that Mügge's numbers cover too wide a period, but conversely, Choenni's estimate is on the low side because he did not include the period of family reunification (Photo 10.4).

One may wonder what impact the flight had on Suriname and especially on agriculture. The figures do not tell a unified story. In time series over the 1970s, De Rooij and Van Schaaijk (1983) show a steady increase in the number of jobs, except in agriculture, where in the years from 1973 to 1979 employment declined from 16,000 to 13,900 places. But the value of rice exports rose from Sf 8.6 million to Sf

Photo 10.4 Surinamese on arrival at Schiphol Airport in 1973

36 million between 1972 and 1976. Conversely, Mhango (1980, p. 74) reported a decline in imports in the period 1974–1976 (especially of raw materials and consumer goods) and a corresponding decline in export earnings (p. 94), reflecting a slump in foreign trade.

Yet there was no decline in real gross domestic product (GDP). Instead, figures by Chin and Buddingh (1987, pp. 127 and 128) showed an increase in agricultural production from 8% to 10% over the years 1971–1975 and an increase in agricultural exports from 6% to 14% over the same period. Hindori (1998, p. 30) presents a graph showing that rice production increased from 145,399 to 174,845 tonnes between 1970 and 1975, declining from 174,845 to 172,500 tonnes from 1975 to 1976 before continuing to grow thereafter. A similar pattern was described by Morenc (1988, p. 240). In short, emigration, which started in late 1973, had no significant impact on rice production. The general picture is of rising production, and this wrangles with the idea that the emigration of at least a fifth of the population is not reflected in the macroeconomic magnitudes.

The question is how the growth can be reconciled with the large demographic decline in the economy. The most likely explanation is that output was borne mainly by the capital-intensive sectors, especially mining (bauxite) and mechanised agriculture. The implication of this explanation is that the hidden unemployment must have been so huge that the economy was hardly affected by the drain of labour. This hidden unemployment must have been mainly in the public sector and small agriculture.

A few authors provide alternative explanations specific to the rice sector. Sital (2002, p. 58) suggests that around independence, emigration created a scarcity of labour. The population declined from 387,525 in 1974 to 354,860 by mid-1980. The labour scarcity had a pull factor on Guyanese workers, estimated at 15,000 people, amounting to about 15% of Suriname's labour force at the time. Morenc (1988, p. 123) also points out that after independence, Suriname was overwhelmed with development projects that began to play a metropolitan function in the region: 'The increase in economic activity in Nickerie and the rest of Suriname, the emergence of large-scale rice cultivation, combined with the economic crisis in Guyana, led to the influx of (agricultural) workers from that country ... These workers, mostly Hindostanis, came from the rural areas and mainly perform seasonal labour.' Singelenberg (1983, p. 30), who conducted an investigation into Guyanese workers in Suriname, often recorded the statement 'Without Guyanese no rice in Nickerie'. Any shortage in the labour market was thus (partly) met by the immigration of Guyanese workers.

By 1980, Hindostani peasants had almost vanished. A handful of large commercial farmers in Nickerie took over 95% of rice production. There was no policy for the benefit of Hindostanis through the rice sector or otherwise. The districts had been transformed into urban areas. Small-scale agriculture consisted predominantly of horticulture around the city and the main towns of the districts, along with subsistence rice cultivation (besides fruits and vegetables) and commercial fishing. The Hindostani community, which had previously found its identity as an ethnic community in agriculture, and established its home there too, hardly existed anymore. It had migrated to the city and the Netherlands or still lived in the districts but had become urbanised. With that, the Hindostani peasants were transformed into urban dwellers.

10.6 Conclusions

The transformation from peasant to urban dweller appears to have been a more complex affair in Suriname than presented in the literature. Internationally, this transformation has been presented as a surplus population in agriculture that should have been absorbed by industrialisation. Because a sizeable industrialisation did not occur in most developing countries, the urbanisation led to ghettos where the poor were concentrated. In Suriname, such ghettoisation did not occur despite the absence of industrialisation. This was probably due to four circumstances: the relatively small population, the inclusion of jobseekers in the public sector, the decreasing discrepancies between city and district, and emigration to the Netherlands.

The transformation concerned three categories of peasants that also expressed three different processes. The first was the poverty that forced many farmers to migrate to the city or be condemned for life to a poverty-stricken existence in the district. These people, so-called 'traditional peasants', were the socially needy whose children migrated or acquired a job in a sector outside agriculture. Rice farming became the responsibility of these left-behind peasants. Second, the marginalised farmers, who seem to be prosperous at face value, had to acquire supplementary income elsewhere. Farmers in Nickerie, but not only there, are examples of this category. The third category comprised farmers who saw in emigration to the

Netherlands a godsend. They traded not only the sector of their livelihood but also the country in which they lived. The combination of the above categories and processes led to a quantitative and qualitative decline of the Girmitiya peasantry in Suriname.

Note that the concept of mobilities encompasses spatially, economic, political and cultural transformation. It comprised issues such as investment and credit, land and labour productivity, crops grown, price formation, marketing and the like. The mobilities were also affected by the power differences between Paramaribo and the districts. Because all facilities were (and are) concentrated in Paramaribo, that area exerted a magnetic attraction on district residents. Paramaribo thus became a frame of reference, a promise that attracted many young district residents.

Cultural factors are probably the most underestimated forces in mobilities. Where economic and political forces are known and recognised in wealth creation, culture is often considered a residual factor. But its relevance can also be reversed. Culture, in the form of community building, could provide a home and stabilise the peasants. Governor Kielstra had promulgated this insight. Later that policy was continued in the form of centre plans, but it was minimally designed and did not provide the amenities that ensured farmers and their children would have a livelihood. Also underestimated is the effect of secondary education that was Western in style. This prepared the peasant children for a life outside agriculture, either in the city or in the Netherlands. In other words, they lost their agricultural capital but acquired new forms of human capital that could be cashed in, either in other sectors or in other countries. The Girmitiyas had become an urban community.

References

Alexander C, Chatterji J, Jalais A (2018) The Bengal diaspora. Rethinking Muslim migration. Routledge, London

Beck S (1924) Landbouwbedrijfsvorm en landbouwopleiding. In: Beck S (ed) Surinaamsche Problemen. Kersten, Paramaribo, pp 1–44

Bovenkerk F (1975) Emigratie uit Suriname. Antropologisch-Sociologisch Centrum, Amsterdam

Burger JW (1928) Vergelijkend Overzicht van de Immigratie en Blijvende Vestiging van Javanen en Brits-Indiërs in Suriname. De Economist 77(1):422–445

Chin H, Buddingh H (1987) Surinam. Politics, economics and society. Francis Printers, London/NY

Choenni C (2016) Hindostaanse contractarbeiders 1873–1920. Van India naar de plantages in Suriname. LM Publishers, Volendam

de Houven V, van Oordt WJ (1900) Een en ander over kleinen landbouw in de kolonie Suriname. De Economist 49(2):599–615

De Rooij W, Van Schaaijk M (1983) Ekonomische Ontwikkeling van Suriname. Vaco, Paramaribo

De Vries M (1965) De Landbouwers van Suriname. Planbureau, Paramaribo

De Wit TPM (1960) The Wageningen rice project: a study on the development of a mechanized rice farming in the wet Tropics. Doctoral dissertation, Wageningen

DLS (1969) Derde Landbouwtelling Suriname, november 1969. Ministerie van LVV/ABS, Paramaribo

Gobardhan L (2001) Onderwijs Als Sleutel Tot Maatschappelijke Vooruitgang. Een Taal- En Onderwijsgeschiedenis Van Suriname, 1651–1975. Doctoral dissertation, University of Leiden

Gowricharn RS (2020) Multiple homemaking. The ethnic condition in the Indian diaspora. Routledge, London
Hasselbach O (1966) The introduction of lodging resistant rice varieties in peasant agriculture. De Surinaamse Landbouw 14:45–55
Heybroek F (1978) Evaluatie en Inventarisatie van het project Santo Boma. Ministerie van LVV, Dienst Landbouw–Economische Zaken, Paramaribo
Hindori D (1998) Geschiedenis van onze landbouw. In: Mitrasingh B, Harpal M (eds) Hindostanen. Van contractarbeiders tot Surinamers. Stichting Hindostaanse Immigratie, Paramaribo, pp 19–39
Kool R (1964) Agricultural planning in Suriname 1950–1960: An evaluation. Wageningen, Veenman & Zonen
Kruijer G (1949) Landbouw-Cooperatie onder de Aziatische bevolkingsgroepen van Suriname en Brits-Guyana. Nieuwe West-Indische Gids 30(1):209–235
Kruijer G (1951) Urbanisme in Suriname. Tijdschrift van het Koninklijk Nederlands Aardrijkskundig Genootschap, jrg 58:31–63
Lewis WA (1954) Economic development with unlimited supplies of labour. In: Agarwala AN, Singh SP (eds) The economics of underdevelopment. Oxford University Press, London, pp 400–449
Leys JJ (1921) De Kleine Landbouwers en het Klein Landbouwbedrijf. West-Indië, Landbouwkundig Tijdschrift voor Suriname en Curaçao 6(1):31–44
Luning H (1966a) Bedrijfsgrootte, werkgelegenheid en agrarische planning. De Surinaamse Landbouw 14(1&2):36–44
Luning H (1966b) Enkele kanttekeningen aangaande de taak van de landbouw bij de ekonomische ontwikkeling van Suriname. De Surinaamse Landbouw 14(4):117–135
Luning H (1967) Economic aspects of low labour-income farming. Doctoral dissertation, University Wageningen
Luning H (1968) Transforming traditional rice farming: The Surinamese experience. Neth J Agric Sci 16:253–263
Luning H, Sital P (1979) The economic transformation of small holder rice farming in Surinam. In: Cross M, Marks A (eds) Peasants, plantations and rural communities in the Caribbean. Leiden, KITLV, pp 193–219
Meel P (1994) Verbroederingspolitiek en nationalisme: het dekolonisatievraagstuk in de Surinaamse politiek. Bijdragen en Mededelingen betreffende de geschiedenis der Nederlanden 109(4):638–659
Menke J, Sno I (2016) Ras en etniciteit in volkstellingen van Suriname. In: Mencke J, Sno I (eds) Mozaïek van het Surinaamse volk. Volkstellingen in demografisch, economisch en sociaal perspectief. UvS/ABS, Paramaribo, pp 76–94
Mhango B (1980) Gap analysis applied to the Surinamese economy. Doctoral dissertation, University van Suriname
Morenc J (1988) Surinaamse Kleine Landbouw en Landbouwbeleid: Een Structurele Analyse. Doctoral dissertation, Radboud University of Nijmegen
Mügge L (2012) Ideologies of nationhood in sending-state transnationalism: comparing Surinam and Turkey. Ethnicities 13:338
Ramadhin H (2014) De historische ontwikkeling van de rijstbouw. Het verleden, het heden en de toekomst. Published in-House
Ramsoedh H (2018) Surinaams Onbehagen. Een sociale en politieke geschiedenis van Suriname 1865–2015. Verloren, Hilversum
Rest CPM (1974) Het departement van Landbouw, Veeteelt en Visserij. Onderzoeksproject Sociale Ontwikkelingsstrategie. University of Amsterdam, Amsterdam
Schiller N, Salazar N (2013) Regimes of mobility across the globe. J Ethn Migr Stud 39(2):183–200
Singelenberg R (1983) Guyanese gastarbeid in Suriname. Instituut voor Culturele Antropologie, Utrecht

Sital JT (2002) How Suriname came to its own green revolution. A study of change in small-holder farming in the Nickerie district, 1900–1986. University of Wageningen, Wageningen
Speckmann J (1965) Marriage and kinship among the Indians in Surinam. Van Gorcum, Assen
TJP (1954) Tienjarenplan voor Suriname. Herziene opstelling door de Stichting Planbureau Suriname. Planbureau, Paramaribo
TLS (1959) Tweede Landbouwtelling Suriname, november 1959. Ministerie van LVV, in samenwerking met ABS, Paramaribo
Uitbuiting (1970) De Uitbuiting van Suriname. SUN, Nijmegen
Van Dusseldorp DBWM (1963a) Het Centraplan voor Suriname: Some aspects of town and country planning in a developing country. Wageningen, Veenman & Zonen
Van Dusseldorp DBWM (1963b) Geografische Mobiliteit en de ontwikkeling van Suriname. Bijdragen tot de Taal-, Land-, en Volkenkunde 199(1):18–55
Van Lier R (1977) Samenleving in een grensgebied: Een sociaal–historische studie van Suriname. Emmering, Amsterdam
Zielhuis L, Girdhari D (1973) Migratie uit Suriname. Afdeling Wetenschappelijk Onderzoek en Planning, Paramaribo

Chapter 11
Conclusions: Transformations of the Girmitiya Peasantry

Abstract This concluding chapter summarises the main findings of the book, specifically that the Girmitiyas were transformed from peasants in British India to labourers on Surinamese plantations, from labourers to peasants and from peasants to urban urbanites. It underscores that the transformations were enabled by the possession or loss of agrarian human capital. Furthermore, it emphasised that all generations have paid a heavy price as the achievements of later Girmitiya entrepreneurs, politicians, academics and other socially successful people rest on their fragile shoulders. Without the sacrifices of these generations, it was not possible to build an ethnic community, develop internal networks, celebrate role models, accumulate various forms of capital and receive several forms of support. The fate of the plantation workers and the peasants who struggled, suffered and perished anonymously cannot simply be termed 'progress' as the sacrifices of several generations of British Indian peasants were too great. The history of the Girmitiya peasants reveals as many reasons to rethink the past as to celebrate.

Keywords Transformations · Sacrifices · Progress · Celebration · Rethinking the past

In the historiography of British Indians and their descendants in Suriname, it is striking that the intellectual effort is focused on indentured labourers. Their history commences on the plantations, with a brief run-up in the recruitment period in British India. The story of Surinamese Girmitiyas, locally known as Hindostanis, is presented as workers who were brought in in appalling conditions and managed to survive on the plantations. Then almost a century is skipped, namely the period when several generations became (part-time) peasants. Next, we find Girmitiyas in the city of Paramaribo. What completely falls away in the description is the acknowledgement that the first generation (who were both workers and peasants) and the later generations of peasants laid the cultural, economic and political foundation for social progress, and that historically they were subject to three transformations and paid a high price for their achievements.

R. Gowricharn, *The Girmitiya Peasants in Suriname*,
https://doi.org/10.1007/978-3-031-67961-2_11

This book addresses a composite problem and aims to achieve three objectives, namely: to explain how the ethnic group formation was linked to the economic position of British Indians; to conduct an analysis of the three transformations in order to show the continuity and ruptures; and to detail the sacrifices made by the indentured labourers and the peasants.

The book has adopted as its central working hypothesis the fact that the Girmitiyas were not unskilled labourers as is assumed in the scholarly literature but rather possessed agrarian human capital, i.e. knowledge of the cultivation of sugarcane and rice and many other commodities as well as the handling of labour instruments and knowledge of vegetation, climate and soil. The hypothesis presupposes that there is a link between group culture and the agrarian economy.

In the Caribbean literature, British Indian immigrants were erroneously approached as replacements for slaves. Colonial sources suggest that this was mainly the concern of the government and planters and later historians have adopted this perspective. This approach misrepresents three important elements of proletarianisation: first, that recruited labourers possessed agrarian human capital; second, that workers had vegetable gardens and sometimes raised livestock and poultry and thus were able to maintain their agricultural knowledge and skills; and third, that agricultural human capital was not an individual asset but a collective form of capital, a kind of communal capital. To mobilise and utilise this capital, the Girmitiyas needed to establish an ethnic or a moral community. The productive forces, consisting of labour, knowledge and skills, agricultural materials and moral ties, were thus embedded in the ethnic community. That is to say, the British Indians harnessed productive forces embedded in their group culture. That was their major asset enabling them to survive plantation and peasant life.

The three transformations were not isolated phenomena but had a continuous line of immigration, colonisation and urbanisation. The first two transformations were enabled by agrarian productive forces, while the third came about because this capital had become obsolete. All three transformations thus had more or less the same causal forces. However, several authors have pointed out that the group of workers was heterogeneous in terms of composition and experience. But the implications of this observation have not been drawn either. These differences render an uneven start in agriculture.

The second transformation, from labourer to peasant, or peasantisation, has received very little attention in either the Surinamese or the international literature on peasants, including that of the Indian diaspora. What has been written about peasants was conceptualised as a continuation of the labour supply for the benefit of the plantations. Peasants were usually detailed as part-timers, and in some countries, such as Guyana, Trinidad and Mauritius, they were allotted land to grow food for the local population. But in essence, they were conceptualised as extensions of the plantations.

Production conditions varied socially and regionally. Because of differences in individual health, knowledge of agriculture, whether or not one had a family, the size of the family, the size and quality of the land, the availability of savings or loans, the distance of the plot from markets, the possibility of transport, price trends, the differences in diligence and thrift and the starting conditions in agriculture were also unequal, an element that has been little addressed. These unequal conditions determined the start and performance of the peasants. But where the plantation still offered some protection through labour rights agreements, peasants were at the mercy of nature, the market and the authorities. Since the colonial authorities were more concerned with macroeconomic issues such as exports and government finances, the fate of the peasants represented a blind spot. This blindness also characterised later researchers.

Although family formation and an ethnic community got off the ground, not all group members could make full use of their agricultural human capital. Plots were sometimes too small, labour productivity low, location with respect to markets unfavourable, infrastructure insufficient, a family relationship sometimes absent or weak (sometimes due to the departure of children), health often fragile, investment opportunities limited and the population in local ethnic residential centres too small. Despite possessing agrarian human capital and being part of a moral community that enabled the mobilisation of additional productive capacity, many peasants had no prospect of a viable future in smallholder agriculture.

These conditions led to a differentiation of occupations early in history. Most notable was the development of a social stratum of traders (in money, poultry and livestock products, vegetables and rice), of service providers (owners of rice paddy mills and of transporters) or peasants' children who aspired to white-collar jobs. It was an early flight from agriculture. Attempts to curb this urbanisation by creating village communities and residential centres had little effect. The main reason for this failure was not that household incomes were inadequate but that amenities in the districts, including social activities education, employment and recreation, were minimally present. In policy circles these preconditions for a sustainable peasant life were not conceptualised as independent needs, but as instrumental conditions for promoting economic performance and inhibiting urbanisation.

Thus, the transformation to urban dwellers had already begun before World War II. Post-war development planning did not signal a break with previous policies. Typical features of the pre-war peasants' perspective, namely a focus on family farms, the duality of family incomes, the mechanisation of rice farming and production for export, persisted to a large degree. The post-war policy focused mainly on the mechanisation of medium-sized farms under the auspices of the SML and the Prins Bernhardpolder. Nevertheless, family farms continued to need supplementary income from outside agriculture.

Mechanisation did not produce a small productive class of farmers, nor did it lead to the financial independence of family farms. The economic and technological perspective remained dominant, not only at the family level but also at the macro level: rice was seen as an export product to generate foreign exchange and increase state revenues. Other dimensions of peasant life that had previously received the attention of agronomists and governors, such as social cohesion and ruralisation, disappeared into the background. The involvement of the VHP did not alter this neglect.

The third transformation, basically the demise of the Girmitiya peasantry, found its completion in the quantitative decline and qualitative change of urban dwellers. This decline was not only geographical but also social and cultural. It comprised the migration from the district to the city, the creation of new jobs in the districts as part of development planning and emigration to the Netherlands. The migration was a push from agriculture as well as a pull to the city. They constituted a comprehensive transformation, rendered by poverty, a lack of prospects in traditional agriculture, Western education and the higher status of urbanisation.

The third goal aimed to detail the sacrifices or the price the Girmitiya peasant paid when developing the moral community and the ethnic institutions so crucial for their survival. That price was mainly paid in three forms. One was the mortality and poverty from which several generations suffered. You will recall that almost one in six immigrants died in Suriname, and the proportion that eventually settled in Suriname was about one-third of the number of immigrants that landed. Considering the short life span in that period, the century must have comprised several generations. Apart from a brief period between the world wars, when the government thought peasants had some significance for exports, hardly any attention was given to them.

The second form is the 'silent suffering' of the loss of a country, kith and kin. This loss accompanied the Girmitiyas throughout their whole life. It was transferred by means of stories of deception and bad luck and the creation of a heritage according to which generations of Girmitiyas had to cope with extremely harsh social and economic conditions.

The third form of the price paid consisted of the migration of children to the city (for work, employment or personal reasons), leaving the parents behind. As time went on, the rural areas became inhabited with old and sick people, abandoned by the children. Those left behind lacked the required support from kin, and their agrarian human capital was rendered useless by the modernisation of agriculture. Without exaggeration, one may say that several generations perished anonymously in the Surinamese countryside.

Despite indications of socio-economic progress among Girmitiyas, poverty remained a persistent feature of peasantry. From the first signals in the first decades of the twentieth century, this issue was often raised by individual agronomists. In every decade, there was at least one author who drew attention to the issue and increasingly also provided numerical support for it. Even peasants in Nickerie, the centre of mechanisation, were forced to earn supplementary income outside agriculture or live out their lives in poverty. Historically, poverty and loneliness remained the peasants' most loyal companions, in many ways far more loyal than their own children.

Girmitiya communities celebrate themselves for their achievements, implying that the left-behinds in India did not 'make' it. They often overlook the fact that their progress is thanks to the various generations of peasants on whose fragile shoulders later entrepreneurs, politicians, academics and other socially successful people rest. Without the sacrifices of these generations, it was not possible to build an ethnic community, develop internal networks, celebrate role models, accumulate various forms of capital and receive other forms of support. The attitude shows a high degree of disdain and a complete lack of understanding of history to think that such developments took place not in India but only in Girmitiya societies. Moreover, all countries have experienced such transformations and forms of mobility. The fate of the plantation workers and the peasants who struggled and suffered and perished anonymously cannot simply be termed 'progress' for this reason. The sacrifices of several generations of British Indian peasants were too great. The immigration of Girmitiyas therefore provides as many reasons to rethink the past as to celebrate.

Annexes

Annex 2.1: Immigrants by Year of Arrival

1873	2541
1874	1413
1875	
1876	
1877	327
1878	700
1879	
1880	785
1881	428
1882	471
1883	902
1884	2134
1885	532
1886	
1887	360
1888	
1889	1747
1890	1275
1891	608
1892	1240
1893	1093
1894	1211
1895	1623
1896	1167
1897	
1898	616

(continued)

R. Gowricharn, *The Girmitiya Peasants in Suriname*,
https://doi.org/10.1007/978-3-031-67961-2

1899	615
1900	
1901	
1902	1331
1903	654
1904	253
1905	173
1906	632
1907	1674
1908	1866
1909	1424
1910	476
1911	
1912	1200
1913	945
1914	1567
1915	
1916	303

Source: Adapted from De Klerk (1953: 71–73)

Annex 2.2: Female Immigrants to Suriname, 1881–1916

Year	Total immigrants	Of which women	Women as %
1881	496	156	31
1882	754	117	16
1883	1304	374	29
1884	1476	429	29
1886	305	88	29
1888	429	123	29
1889	1013	290	29
1890	1067	307	29
1891	511	148	29
1892	1067	304	28
1893	1665	520	31
1894	1086	304	28
1895	1411	416	29
1896	1096	304	28
1897	597	173	29
1898	562	163	29
1899	560	154	28
1901	678	218	32

(continued)

Year	Total immigrants	Of which women	Women as %
1902	1322	346	26
1904	231	58	25
1905	171	44	26
1906	1270	320	25
1907	1091	265	26
1908	1833	486	27
1909	1906	503	26
1912	1216	309	25
1913	1773	461	26
1914	748	188	25
1916	395	81	21
Total	27,932	7649	27

Source: Bhagwanbali (1996: 96)

Annex 2.3: Returned Migrants 1878–1921

Year	Men (1)	Women (2)	Children (3)	Total (4)	2 as % of 4
1878	265	114	107	486	23
1879	207	80	46	333	24
1884	298	131	108	537	24
1886	331	136	156	623	22
1887	290	98	92	480	20
1889	237	142	148	527	27
1890	333	116	121	570	20
1891	360	122	119	601	20
1895	339	142	124	605	23
1897	131	49	33	213	23
1898	434	157	152	743	21
1899	157	36	21	214	17
1900	142	49	38	229	21
1903	236	50	32	318	16
1904	189	41	35	265	15
1905	485	134	115	734	18
1907	182	29	20	231	13
1909	371	76	54	501	15
1911	418	105	80	603	17
1913	488	106	74	668	16
1920	586	205	183	974	21
1921	513	199	174	886	22
Total	6992	2317	2032	11,341	20

Calculated from Choenni (2016: 509)

Annex 3.1: Ratio of Men to Women by Ethnic Group, 1922–1970

Year	Creoles	Hindostanis	Javanese	Country
1922	89.2	127.3	138.9	101
1923	88.1	129	143	101.4
1924	87.3	126.6	141.8	101.5
1925	88.5	124.8	135.5	101.1
1926	86.8	123.3	133.3	100.4
1927	86.6	121.9	131.7	100.1
1928	85.8	120.8	132.9	100.4
1929	85.6	118.7	130.9	99.7
1930	85.5	118.1	129.7	99.4
1931	85.6	118	128.2	99.3
1932	85.6	117.7	126.7	99.1
1933	85.6	117.8	124.7	99.1
1934	86.1	117.4	124.3	99.1
1935	86.4	115.4	123.3	98.7
1936	86.5	114.6	121.6	98.4
1937	87.8	111.5	120	97.7
1938	87.3	110.4	119.5	96.9
1939	88	110.8	121.9	98.4
1940	89.9	108.5	119.2	98.4
1941	89.4	108	118.7	98
1942	89.6	107.4	117.6	97.8
1943	89.9	111	116	97.9
1944	90.7	106.6	115.5	96.6
1945	89.6	107.9	117.9	96.9
1946	89.7	107.9	117.3	97.2
1947	90	107.4	115.6	96.9
1948	90.8	106	114.9	97
1949	93	107	113.9	98.5
1950	86.7	105.5	108.2	98.3
1951	87.1	104.9	107.7	98.2
1952	87.8	104.6	107.4	98.3
1953	88.6	104.3	106.4	98.3
1954	89.1	104	106	98.3
1955	89.7	103.8	105.2	98.3
1956	90.1	103.4	104.9	98.3
1957	90.5	102.8	104.4	98.2
1958	90.8	102.2	104.2	98
1959	91.3	102.1	103.8	98.1

(continued)

Year	Creoles	Hindostanis	Javanese	Country
1960	91.7	101.9	103.3	98.1
1961	–	–	–	98.2
1962	–	–	–	98.1
1963	95.2	101.9	104.3	99.8
1964	95.3	101.3	103.4	99.5
1965	95.5	101	103.3	99.4
1966	95.5	101	102.7	99.4
1967	95.9	101.1	102.5	99.5
1968	96.1	100.9	102.1	99.4
1969	96.3	100.8	101.8	99.4
1970	96.4	100.4	101.5	99.3

Source: Lamur (1973: 178–79)

Annex 3.2: Average Wages of Hindostani Migrants by Work Task, 1875–1891

Year	Able-bodied	Not able-bodied
1875	**0.66**	0.56
1876	0.67	0.64
1877	0.69	0.60
1878	0.70	0.64
1879	0.70	0.63
1880	0.69	0.60
1881	0.69	**0.43**
1882	0.73	0.65
1883	**0.74**	0.63
1884	0.72	0.62
1885	0.72	0.61
1886	0.72	0.62
1887	0.71	0.62
1888	0.71	0.61
1889	0.69	0.61
1890	0.70	0.59
1891	0.71	**0.70**
Average	0.70	0.61

Source: Bhagwanbali (2010: 157)
Extreme values are in bold

Annex 3.3: Wages Paid to Hindostani Contract Workers in Dutch Guilders

Year	Number of men	Total wages	Average per man	Number of women	Total wages	Average per woman
1875	1707	181,979.52	0.56	660	46,738.41	0.47
1876	1652	190,997.98	0.61	629	48,232.17	0.51
1877	1639	193,889.77	0.62	647	51,623.12	0.53
1878	1792	216,980.48	0.64	674	56,820.93	0.56
1879	1633	219,929.64	0.71	563	48,456.44	0.57
1880	1839	258,427.18	0.74	625	51,070.01	0.54
1881	2345	333,459.50	0.75	800	70,159.97	0.58
1882	2586	368,089.04	0.75	958	93,941	0.65
1883	2619	387,358.34	0.78	955	83,642.41	0.58
1884	3045	429,936.86	0.74	1084	103,969.69[a]	0.64
1885	3595	485,052.07	0.71	1443	103,969.69[a]	0.48
1886	3303	464,688.13	0.74	1128	94,868.34	0.56
1887	3245	453,577.86	0.74	1147	89,307.49	0.52
1888	2816	397,366.72	0.74	955	75,481	0.53
1889	2583	324,897.82	0.66	813	64,120.57	0.53
1890	2828	359,424.23	0.67	816	57,910.89	0.47
1891	3306	442,070.43	0.70	1002	71,551.54	0.48
Average			0.70			0.54

Source: Bhagwanbali (2010), Annex 3
[a]Possible misstatement

Annex 3.4: Values Taken by Departing British Indians, 1878–1921

Year	Number of returnees (1)	Money value taken (2)	Jewellery and coins (3)	Total values (2 + 3)	Per capita value (2 + 3)/1
1878	436	44,100.25	2203.65	46,303.9	106.20
1879	333	23,110.85	2350	25,460.85	76.46
1884	537	115,387.58	9218	124,605.58	232.04
1886	623	78,287.43	8900	87,187.43	139.95
1887	480	37,908.02	10,333	48,241.02	100.50
1889	527	95,126.35	20,045	115,171.35	218.54
1890	570	71,675.18	10,746	82,421.18	144.60
1891	601	86,034.1	16,563	102,597.1	170.71
.1895	605	89,687.37	15,501.35	105,188.72	173.87
1897	213	26,837.37	3697.6	30,534.97	143.36
1898	743	91,588.72	14,750	106,338.72	143.12

(continued)

Year	Number of returnees (1)	Money value taken (2)	Jewellery and coins (3)	Total values (2 + 3)	Per capita value (2 + 3)/1
1899	214	35,933.77	2501.5	38,435.27	179.60
1900	229	40,600.3	5572.25	46,172.55	201.63
1903	318	48,245.54	5673	53,918.54	169.56
1904	265	27,787.43	5099	32,886.43	124.10
1905	734	80,084.71	11,433.5	91,518.21	124.68
1907	231	33.311	4322.5	37,633.5	162.92
1909	501	38,700.33	4026.5	42,726.83	85.28
1911	603	57,995.86	10,800	68,795.86	114.09
1913	668	70,323.45	0	70,323.45	105.27
1920	974	187,874.71	29,625	217,499.71	223.31
1921	886	201,113	25,000	226,113	255.21

Source: Choenni (2016: 509)

Annex 4.1: Ten Plantations with the Largest Number of Hindostani Contract Workers

Name	1875	1880	1885	1890	1900	1915	1920
Alkmaar	95	150					
Alliance	170	197	215	287	433[b]	360	67
Burnside	95	178					
Catherina Sophia		132			54		
De Eendracht	96				62		
De Maasstroom							27
De Resolutie	461	705	261				
Dordrecht					50	151	21
Goudmijn				230			
Hazard			305	191		91	62
Jagtlust				151	221	106	
Kroonenburg	161						
Leasowes and Sara	120						
Mariënbosch							20
Mariënburg (and Zoelen 1900)			422	576	1101[b]	275	61
Meerzorg	95	131					
Mon souci							17
Peperpot						151	30
Rust en Werk	184	206	285	272	166[c]	105	
Slootwijk						88	
Visserszorg			186				
Voorburg		180	325	211			
Vreeland	106						

(continued)

Name	1875	1880	1885	1890	1900	1915	1920
Waterland					51		
Waterloo	110	210	396	228	393[a]	239	58
Wederzorg					113		
Zoelen			439	689		292	64
Zorg en hoop	108	168	180				
Total	1801	2257	3014	2835	551	1858	427
Average per plantation	150	226	301	354	42	186	43
Range per year	95–461	131–701	180–439	151–689	51–551	91–360	17–67

Source: Adapted from Fokken 2018: 331–335.
Note: Not all columns add up to ten (plantations).
[a]Waterloo together with Nursery
[b]Alternately classified as Matapica, Cottica
[c]Together with Zoelen
[d]Before 1900 classified as Beneden-Commewijne

Annex 4.2: Number of Creole and Asian (British Indian and Javanese) Workers on the Plantations, 1862–1910

	Creoles		Immigrants		Total
Year	Number	In %	Number	In %	
1862	16,824	98	297	2	17,121
1872	13,390	86	2214	14	15,604
1873	10,604	71	4229	29	14,903
1874	9870	67	4851	33	14,721
1875	9127	70	4002	30	13,129
1876	8864	71	3605	29	12,469
1877	7730	68	3666	32	11,369
1878	7370	68	3666	32	10,860
1879	7119	70	3069	30	10,180
1880	6889	69	3091	31	9918
1881	6907	67	3365	33	10,272
1882	6435	65	3538	35	9973
1883	6340	67	3174	33	9514
1884	5223	50	5155	50	10,378
1885	5282	52	4898	48	10,180
1886	4409	48	4723	52	9132
1887	4108	48	4385	52	8439
1888	4775	59	3253	41	8028
1889	5040	60	3325	40	8365
1890	4838	53	4315	47	9153
1891	4559	49	4718	51	9277
1892	4662	47	5208	53	9870
1893	4516	45	5417	55	9933

(continued)

Year	Creoles Number	In %	Immigrants Number	In %	Total
1894	5256	48	5674	52	10,930
1895	4941	45	5969	55	10,910
1896	5507	48	5994	52	11,501
1897	6017	49	6228	51	12,305
1900	5501	48	5850	52	11,351
1902	5801	44	7422	56	13,223
1910	1737	10	14,813	90	16,551

Source: Hoefte (1998: 129)

Annex 5.1: Exports of Sugar, Coffee, Cotton and Cocoa, 1730–1862 (Sugar in Metric Tonnes, the Other Products in Pounds)

Year	Sugar	Coffee	Cotton	Cocoa
1730	10,600	503,667	–	–
1740	9470	4,971,246	3834	77,853
1750	11,980	3,536,339	114	338,884
1760	8300	10,206,487	1292	128,482
1770	5740	230,547	–	168,260
1780	6180	11,750,000	650,800	570,350
1790	8300	14,856,275	943,580	517,464
1795	12,142	11,193,690	1,223,170	350,444
1805	5865 (72)	6,100,755 (64)	5105 (-)	273,488 (68)
1810	8500 (105)	6,504,596 (68)	7402 (-)	88,245 (22)
1816	5500 (68)	4,843,120 (51)	851,620 (91)	90,695 (23)
1818	9400 (116)	8,422,178 (88)	1,339,615 (143)	97,489 (24)
1820	9800 (121)	4,815,037 (51)	1,328,243 (141)	8715 (2)
1823	10,500 (130)	6847, 631 (72)	1,889,402 (201)	68,960 (17)
1825	11,900 (147)	4,196,575 (44)	2,329,607 (248)	116,105 (29)
1828	13,600 (168)	5,557,601 (58)	1,586,451 (169)	129,254 (32)
1830	16,500 (204)	6,046,102 (64)	2,033,304 (216)	18,455 (5)
1835	18,900 (234)	2,152,550 (23)	1,040,307 (111)	31,070 (8)
1840	17,200 (213)	3,461,902 (36)	1,993,256 (212)	182,984 (46)
1845	14,900 (184)	1,708,479 (18)	840,505 (89)	104,295 (26)
1850	15,600 (193)	723,514 (8)	939,774 (100)	167,511 (42)
1855	17,400 (215)	528,514 (6)	936,181 (100)	260,617 (65)
1859	14,000 (173)	529,581 (6)	537,214 (57)	459,917 (115)
1862	16,400 (208)	364,298 (4)	378,607 (40)	711.799 (77)

Source: Panday (1959: 19, 20, 78). Index numbers added
- = less than 1 or not available

Annex 5.2: Five-Year Averages of the Value of Exports, 1866–1935, in Percent

Period	Sugar	Coffee	Cocoa	Gold	Rubber	Bauxite	Exports in *Sf* 1000
1866–70	71	–	10	–	–	–	2722
1871–75	73	–	14	–	–	–	2870
1876–80	60	–	23	12	–	–	3051
1881–85	35	–	29	29	–	–	3681
1886–90	19	–	34	36	–	–	3538
1891–95	24	–	38	30	5	–	4774
1896–1900	23	–	42	23	6	–	5181
1901–05	28	1	31	24	13	–	4380
1906–10	29	1	18	24	23	–	6315
1911–15	32	3	3	15	33	–	7715
1916–20	26	13	17	12	28	–	8036
1921–25	35	18	7	6	26	5	7596
1926–30	14	25	1	2	16	22	9509
1931–35	9	20	–	9	6	35	4535

Source: Colonial Reports and Suriname Report, respectively: statistical annexes
- < 1 or unknown

Annex 5.3: Overview of Settlements and Polders, 1862–1957, by District

Name	District	Year	Size in ha	Hectare
Paradise (s)	Nickerie	1898	286	A
Nw Waldeck (s)	Nickerie	1902	341	A
Sawmillkreek (p)	Nickerie	1909	1295	B
Van Drimmelen (p)	Nickerie	1915	1010	B
Boonacker (p)	Nickerie	1917	173	B
Corantijn (p)	Nickerie	1920	1000	B
Henar (p)	Nickerie	1921	?00	B
Hamptoncourt (p)	Nickerie	1921	885	A + B
Clara (p)	Nickerie	?	?	B
Nani (p)	Nickerie	1949	1460	B
Groot Henar (p)	Nickerie	1951	2740	B
Totness (s)	Coronie	1862	350	A
Totnesspolder (p)	Coronie	1921	128	A + B

(continued)

Name	District	Year	Size in ha	Hectare
Mary's Hope (p)	Coronie	1925	429	A + B
Cocospolder (p)	Coronie	1954	360	B
Vesting achter				
Groningen (s)	Saramacca	1914	164	B
Kampong Baroe (p)	Saramacca	?	?	A + B
Tamaredjo (p)	Saramacca	?		B
Tijgerkreek(p)	Saramacca	?	?	B
La Poule (p)	Saramacca	1954	258	A
Tijgerkreek Oost (p)	Saramacca	1957	220	B
Domburg (s)	Suriname	1873	425	A
La rencontre (s)	Suriname	1897	643	A
Laarwijk (s)	Suriname	1901	238	A
Livorno (s)	Suriname	1904	489	A
Pad van Wanica (s)	Suriname	1904	3668	B
Lelydorp (s)	Suriname	1904	2950	B
Saramaccapolder (s)	Suriname	1906	2160	B
Frederiksburg (s)	Suriname	1913	?	A
Meerzorg (s)	Suriname	1915	1238	A
Adrichem (s)	Suriname	?	324	A
Kassabaholo (s)	Suriname	1918	97	B
Weg naar Zee (p)	Suriname	1953	330	B
Uitkijkpolder (p)	Suriname	1954	1500	B
Houttuin (p)	Suriname	1956	885	B
Nw Amsterdam (s)	Commewijne	1875	129	A
Alkmaar (s)	Commewijne	1875	634	A
Hecht & Sterk (s)	Commewijne	1902	322	A
Joh. & Margareta (s)	Commewijne	1903	345	A
Kroonenburg (s)	Commewijne	1903	973	A
Reynsdorp (s)	Commewijne	1916	602	A
Voorburg (p)	Commewijne	1952	127	A
Welgevallen (p)	Commewijne	1954	240	A
Nw Meerzorg (p)	Commewijne	1955	280	A
Mon Plaisier (p)	Commewijne	1956	50	A

Source: Adapted from (Van Dusseldorp 1963a: 12)
s = place of settlement
p = polder
A = plantation
B = virgin land

Annex 6.1: Indices of Quantities of Peasants' Product, 1906–1935; Average for 1905–1910 = 100

On settlement schemes

Year	(1)	(2)	(3)	(4)	(5)	(6)	(7)	(8)
1906	80	363	61	102	121	126	144	116
1910	128	45	118	141	113	89	90	127
1915	134	132	487	213	416	111	84	150
1920	161	412	866	191	571	130	176	205
1925	111	835	1248	234	218	130	342	362
1930	119	1012	2104	92	288	158	522	731
1935	77	1593	1613	167	363	147	286	1291

Outside settlement schemes

Year	(1)	(2)	(3)	(4)	(5)	(6)	(7)	(8)
1906	105	118	93	105	132	128	82	120
1910	78	104	115	114	99	102	93	135
1915	63	150	256	194	102	79	131	153
1920	71	443	621	314	234	78	267	335
1925	41	776	633	335	215	100	181	327
1930	23	375	1365	214	122	55	201	708

On settlement schemes, 100 =	Outside settlement, 100 =
(1) Cocoa = 873.200 kg	Cocoa = 4.129.400 kg
(2) Coffee = 119.200 kg	Coffee = 101.400 kg
(3) Rice = 3.812.600 kg	Rice = 13.379.200 kg
(4) Corn = 273.000 kg	Corn = 5.412.600 kg
(5) Groundnuts = 3.042.400 kg	Groundnuts = 675.200 kg
(6) Bananas = 2.119.800 bunches	Bananas = 2.416.800 bunches
(7) Plantains = 907.200 bunches	Plantains = 561.200 bunches
(8) Coconuts = 85.500 pieces	Coconuts = 5.798.200 pieces

Source: Colonial Reports and Suriname Report, respectively; statistical appendices

Annex 6.2: Plants and Animals Originating in British India

This appendix was compiled by Lal Goerdayal, biologist and farmer's son, based on his own work and an unpublished study by Trinidadian historian Samaroo (2017). The latter study overlaps somewhat with the studies by Mahabir (2004, 2008).

Animals imported into West Indies/Caribbean countries during immigration from India.

- cattle including the cattle breed zebu
- water buffalo
- sheep
- goats
- mongooses
- rhesus monkeys
- birds including peacocks, Indian silverbill (lead mouth/silverbill), red tiger finch, nutmeg finch

Plants introduced into the West Indies/Caribbean during immigration from India.

- mango/manja
- guava
- pomegranate
- ochre
- long beans
- pumpkin
- marijuana
- thorn apple
- rice
- sapotille
- betel nut
- turmeric
- sopropo
- ginger
- neem
- mustard
- black pepper
- onion
- cinnamon
- lowki (long melon)
- curry plant
- cumin
- fennel
- fenugreek (maithi)
- cloves
- slime apple
- lotus
- various spinach varieties
- Indian pea varieties
- various citrus species
- tulsi (basil)
- jackfruit (nangka, cowa)
- nutmeg

- various hibiscus species (angalampus species)
- poovan banana varieties (bacao varieties)
- different types of Indian teak
- bodhi tree
- different teas

Source: Goerdayal (2019), Mahabir (2004, 2008), Samaroo (2017).

Annex 7.1: Production and Export of Rice in 1000 kg and Rice Prices in Cents, 1929–1948

Year	Production	Export	% exported	Domestic price	Export Price	Price ratio
1929	22,211	3208	14.4	nb	nb	nb
1930	20,795	2265	10.9	nb	nb	nb
1931	24,395	2265	9.3	4	10.5	2.6
1932	25,049	2673	10.7	3	7.5	2.5
1933	27,339	1576	5.7	3.5	6	1.7
1934	12,508	1972	15.7	3	4.3	1.4
1935	27,533	–	–	5	12	2.4
1936	34,168	1288	3.8	6	4.1	0.7
1937	35,355	7403	20.4	5.5	5.6	1
1938	37,578	7136	19	5	6	1.2
1939	40,901	6447	15.8	5	5.9	1.2
1940	30,064	1524	5.1	7	7.5	1.1
1941	50,451	1520	3	9	10.8	1.2
1942	40,359	3231	8	7.8	13.8	1.8
1943	32,711	–	–	13	10.1	0.8
1944	29,454	146	0.5	18	19.5	1.1
1945	33,363	156	0.5	20	24.9	1.2
1946	51,463	2043	4	20	25.1	1.3
1947	39,208	2531	6.5	20	29	2.3
1948	58,395	1184	2	20	22.5	1.1

Source: Columns 1 and 2, Surinaamsch Verslag, statistical appendices
Nb = not known
Column 2: K.v.K. (1950), Appendix V
Column 3: Eswin (1956: 35)

Annex 7.2: Total Area and Rice Area in ha, 1931–1948

Year	Total surface	Surface cultivated	Rice area[a]
1931	120,410	23,671	8132

(continued)

Year	Total surface	Surface cultivated	Rice area[a]
1932	100,200	23,520	8350
1933	109,693	20,165	9113
1934	76,849	20,391	4169
1935	76,755	28,645	9177
1936	76,629	21,706	11,389
1937	73,272	22,377	11,785
1938	73,430	23,773	12,526
1939	76,222	24,170	13,634
1940	72,866	25,686	10,021
1941	68,276	24,897	16,817
1942	67,111	23,339	13,453
1943	78,749	21,079	10,904
1944	67,896	14,711	9818
1945	45,262	16,007	11,121
1946	34,181	18,698	17,155
1947	69,943	31,113	13,136
1948	64,234	18,640	19,465

Source: Suriname Report, statistical annexes
[a]Volume of production divided by average yield = 3000 kg/ha

Annex 7.3: Migrant Population (British Indians and Javanese) in Small Farming, by District, 1931–1948

	Nickerie		Coronie		Saramacca		Suriname		Commewijne	
Year	Number	%	Number	%	Number	%	Number	%	Number	%
1931	6617	91	135	4	3863	76	21,143	75	8488[b]	87
1932	6877	95	137	4	3829	74	20,737	75	6787[b]	84
1933	7325	76	138	4	3990	76	23,067	74	13,722	87
1934	7392	90	146	4	4108	77	24,711	74	11,426	83
1935	7603	91	157	5	4177	76	26,212	74	12,659	84
1936	8025	91	193	6	4266	77	27,381	74	11,072	73
1937	8432[a]	91	249	7	4101	77	27,457	69	11,608	77
1938	8468	90	272	8	4267	77	29,546	75	10,625	79
1939	8552	84	239	7	4379	79	32,252	68	12,450	90
1940	9580	70	239	7	4379	79	32,252	72	12,450	90
1941	9099	72	235	5	5583	82	33,100	69	12,780	90
1942	9327	75	192	7	5553	82	34,024	69	11,551	89
1943	9128	72	223	10	5578	82	16,032	68	10,795	91
1944	9834	82	294	9	5592	81	15,109	87	11,748	92

(continued)

	Nickerie		Coronie		Saramacca		Suriname		Commewijne	
1945	10,112	81	320	10	5610	81	15,194	85	11,869	92
1946	10,267	91	381	10	7659	76	17,778	88	12,028	92
1947	10,553	92	340	9	8173	77	17,748	88	12,320	92
1948	10,996	98	370	10	7072[a]	65	18,960[a]	97	12,452[a]	96

Source: Suriname Report, statistical annexes
[a]Estimate
[b]Probable misstatement

Annex 7.4: Rice Production of Smallholders by District in 1000 kg, 1929–1950

Year	Nickerie	Coronie	Saramacca	Suriname	Commewijne	Marowijne
1929	6880	1410	1975	8279	3615	52
1930	6720	930	2009	7215	3834	87
1931	7366	1029	2118	9204	4591	87
1932	7044	1090	2525	9449	4871	71
1933	9648	1127	3691	8937	3829	108
1934	3494	791	1443	5102	1527	152
1935	10,769	1248	3709	8836	2901	70
1936	10,841	1304	5341	11,563	4972	147
1937	14,273	1713	5740	9176	4312	141
1938	13,768	1297	6702	13,264	3354	109
1939	14,698	1150	7149	13,264	4551	90
1940	2724	252	6056	14,803	6116	113
1941	15,043	1600	6464	15,446	6819	118
1942	13,758	606	7357	12,984	5599	55
1943	10,080	762	4781	13,149	3878	62
1944	10,963	860	6201	7889	3468	74
1945	13,730	1068	6118	8539	3835	73
1946	16,259	1523	6717	20,491	6419	58
1947	13,977	1197	6043	13,598	4334	59
1948	16,772	1513	7832	23,686	8536	56
1949	16,500	1047	7470	18,600	6566	21
1950	17,582	1068	6412	18,832	6181	15

Source: Until 1945 Surinaamsch Verslag, statistical annexes; 1945–1950 Abrahamse 1959: 49–50

Annex 7.5: Report Trial Cuts for the Months of September/October 1940 with Skrivimankoti Rice Crop

Place	Yield	Specification per ha
Reynsdorp	3750	Planted in widened snare; patjolled
Bruinendaal	3350	Ploughed
Constantia	2575	Ploughed
Hecht & Sterk	3350	No tillage
Hecht & Sterk	4150	Planted in widened snare
Frederiksburg	5025	Ploughed; densely planted
Kroonenburg	3900	Ploughed
Kroonenburg	2600	No tillage; polder
Kroonenburg	2825	No tillage; polder
Kroonenburg	3700	No tillage; polder
Kroonenburg	2825	No tillage; polder
Kroonenburg	2175	No tillage; polder
Kroonenburg	2525	No tillage; polder
Kroonenburg	2300	No tillage; polder
Frederiksdorp	4725	No tillage; polder
Frederiksdorp	4600	No tillage; polder
Margrita	3775	Ploughed
Margrita	2850	Patjolled
Margrita	3075	No tillage; polder
Margrita	2500	Ploughed; lack of water
Margrita	2875	Ploughed
Rust en Werk	2975	Ploughed
Rust en Werk	2200	No tillage
Rust en Werk	3475	No tillage
Jagtlust	2650	No tillage
Sorgvliet	3250	No tillage
Leliëndaal	3525	No tillage
Alkmaar-Nijd en Spijt	4125	No tillage
Alkmaar-Nijd en Spijt	4125	No tillage
Alkmaar-Nijd en Spijt	3750	Ploughed
Alkmaar-Kampong Sawah	4375	Ploughed
Alkmaar-Tamanredjo	3075	No tillage
Mon Trésor	3175	No tillage
Katwijk	3550	No tillage
Katwijk	4425	No tillage
Katwijk	3275	No tillage
Meerzorg	4375	No tillage
Meerzorg	4500	No tillage
Meerzorg	3200	Ploughed
Peperpot	4550	Ploughed; new land

Source: LEZ (1941: 57 ff)

Annex 7.6: Cattle Stock by District, 1931–1949

Year	Nickerie	Coronie	Saramacca	Suriname	Commewijne
1931	1273	465	1468	11,969	649
1932	1376	490	858	11,262	898
1933	1213	543	1109	13,055	971
1934	1718	548	1242	13,052	984
1935	2003	566	1678	13,061	889
1936	2241	480	1723	14,183	610
1937	2396	496	1717	14,573	900
1938	2529	545	1788	15,599	753
1939	2653	558	1903	15,599	753
1940	2660	581	1960	15,961	961
1941	2942	626	1498	14,256	1484
1942	3304	541	1323	17,554	1906
1943	3168	609	1096	17,978	1906
1944	3812	623	1132	17,978	1906
1945	4424	690	1180	19,060	1968
1946	4925	701	1216	14,356	1978
1947	5842	878	1291	19,962	2090
1948	5491	730[a]	1406[a]	18,945	2182[a]
1949	6470	845	1227	20,191	2718

Source: Suriname Report, statistical annexes
[a]Estimate

Annex 8.1 Production and Export of Rice and Domestic and Foreign Market Prices in Cents per kg, 1948–1960

Year	Production (in tons)	Export (in tons)	Average domestic price	Average export price
1948	58,395	2415	20	23
1949	50,295	9697	19	33
1950	50,429	4245	18	28
1951	57,704	3668	19	25
1952	54,092	8733	19	29
1953	57,640	7444	21	37
1954	66,659	6417	22	29
1955	64,526	11,891	22	24
1956	71,234	14,674	22	23
1957	55,128	11,250	23	24
1958	85,049	10,241	25	24
1959	78,693	11,837	25	23
1960	80,710	17,937	21	21

Source: Compiled from annual reports of LEZ

Annex 8.2: Concentration of Labour Instruments, 1959

Size class in ha	Ox ploughs	Tractor ploughs and harrows	Tractors	Threshing machines
< 0.48	1	2	2	2
0.48–1	18	5	4	3
1–2	198	28	15	2
2–3	208	50	23	3
3–4	266	56	34	2
4–5	199	67	50	2
5–6	137	41	31	2
6–7	121	34	24	1
7–8	74	43	30	4
8–10	89	87	69	8
10–20	129	156	124	27
20	32	220	216	16
Total	1.472	789	622	72

Source: TLS, Second Agricultural Census (1959: 87, 90)

Annex 9.1: Participants in the First General Election, 1949

1. Agrarian Party
2. Christian Social Party
3. Commewijne Vooruit
4. Ladies' Committee
5. International Workers' Union
6. Kaum Tani Persuatan Indonesia
7. Negro Political Party
8. National Party of Suriname
9. Partyless Comvalius, Th.A.C.
   Baank, A.E.
   Honour, J. van.
   Lieuw Kie Song, A.E.
   Lim A Po, F.H.R.
   Sanches, S.E.H.
   Schneiders-Howard, Ms. G.F. Smit, A.L.R.
10. Pergarakan Bangsa Indonesia Suriname
11. Paranam Miners' Union
12. Progressive Suriname People's Party
13. Suriname Farmers' Organisation
14. United Hindostan Party
    Source: (Sedney 2017: 194).

Annex 9.2: Proposed centres and population in 1950 by ethnicity (%)

Place	Level of care	Total population	Hindostanese	Javanenese	Creoles	Others
Nanni	Local	447	96	3	1	0
Corantyn	Local	2485	84	14	2	0
Paradise	Local	2655	58	36	5	1
Nw Nickerie	Regional	8265	43	20	34	3
Groot Henar	Regional	1692	80	9	10	1
Wageningen	Regional	–	–	–	–	–
Novar	Local	1090	–	–	100	–
Hamilton	Local	523	–	2	98	0
Totness/friendship	Regional	2034	2	12	85	1
Coppenamepunt	Regional	8	0	75	0	25
Carl-François	Local	30	30	4	63	3
Huwelijkszorg	Local	854	99	0	1	0
Bombay	Local	402	85	0	15	0
Calcutta	Regional	819	91	6	3	0
Tijgerkreek	Local	1649	36	58	3	3
Tambaredjo	Local	314	35	54	11	0
Prevoyance	Local	580	83	11	6	0
Monkshoop	Local	501	61	30	9	0
Kampong Baroe	Local	1369	21	73	4	2
Groningen	Regional	1347	27	53	13	7
Jarikaba	Local	207	63	33	4	0
Goede Verwachting	Local	91	98	0	2	0
Kuarsan	Local	820	43	54	3	0
Uitkijk	Regional	1075	63	18	19	0
Duisburg	Local	3360	80	15	4	1
Kwatta	Local	4133	70	16	10	4
Meerzorg	Local	3836	61	34	4	1
Livorno	Local	4935	75	9	15	1
Paramaribo	National	–	–	–	–	–
Santo	Local	5717	90	7	3	0
Houttuin	Local	1675	96	1	2	1
Welgedacht	Local	4905	88	2	8	2
Lelydorp	Regional	663	21	60	16	3
Biliton	Local	1945	10	20	61	9
Domburg	Local	3122	41	51	6	2
Laarwijk	Local	717	58	30	9	3
Paranam	Regional	2585	11	50	33	6
Jachtlust	Local	956	28	68	2	2
Mariënburg	Local	2524	18	70	10	2
Alkmaar	Local	2577	34	62	2	2

(continued)

Place	Level of care	Total population	Hindostanese	Javanenese	Creoles	Others
Tamanredjo	Local	1170	5	94	0	1
Orleana	Local	655	63	33	4	0
Margerita	Local	2606	24	72	2	2
Kroonenburg	Local	1524	73	26	1	0
Nw. Amsterdam	Regional	1826	24	52	21	3
Storkoe	Local	1470	64	34	1	1
Slootwijk	Local	230	3	82	15	0
Alliance	Regional	1610	13	76	7	3
Total (exluding Paramaribo		90,416				

Source: Van Dusseldorp (1963a: 74–75)

Annex 10.1: Ageing in Agriculture, 1959–1969, by District

1959

District/age class	< 35 year	35–44 year	45–54 year	55–64 year	65+ year
Paramaribo	20	40	46	26	14
Nickerie	875	571	461	421	217
Coronie	81	121	182	170	151
Saramacca	509	377	375	395	196
Suriname	2026	1867	1645	1208	948
Commewijne	750	495	642	689	494
Marowijne	11	30	33	45	18
Tot. Land	4272	3501	3384	3044	2038

Source: Second Agricultural Census Suriname (1959: 41)

1969

District/age class	< 35 year	35–44 year	45–54 year	55–64 year	65+ year
Paramaribo	5	11	14	8	3
Nickerie	728	100	535	368	252
Coronie	21	89	105	151	161
Saramacca	342	385	297	271	272
Suriname	1575	2082	1717	1182	948
Commewijne	358	517	368	419	606
Marowijne	68	70	75	57	46
Brokopondo	9	4	17	10	–
Tot. Land	3286	4392	3311	2619	2470

Source: Third Agricultural Census Suriname (1969: 31)

Glossary

ahirwá ke nác peasant dance to the tones of nagárá music (a type of music in which drum beating predominates)
apan jath politics in English-speaking Caribbean—policy aimed at own ethnic group
arkatya recruiters of contract workers in British India
bhaiyá, bhái brother
bahini sister
baithak gáná Hindostani folk music
bibit rice seedling, rice seedlings
birahá song, part of the ahirwá ke nác celebrating rural life
biryani Indian dish where the rice is mixed with vegetables and meat
biyá wedding party, getting married
boiti countryside, district
dáhl lentil soup
dantál iron bar for making music
dhanyá coriander
dhol, dholak two-headed, cylindrical drum
dipuá depot
Diwáli festival of lights of the Hindus
Djamun black-purple fruit, similar to cherries
Eid festival among Muslims
Eid-ul-Adha Muslim festival of sacrifice
Eid-ul-fitr sugar festival
frontier society open society offering land to settlers
gánjá marijuana, drugs
gená africana
gánw ké palwár village relationship, interrelationships in a village
ghughari fried chickpeas

R. Gowricharn, *The Girmitiya Peasants in Suriname*, https://doi.org/10.1007/978-3-031-67961-2

hawan fire sacrifice, Hindu prayer service
Holi powder festival among Hindus
jackfruit fruit of the bread tree, used as stuffing and vegetable like potato
jaháji companion British Indian immigrants who came together on a ship to Suriname
jajmáni system in Suriname a social and religious network of households in which the pandit was the lynchpin
jhandi flag at a religious ceremony
joint family housing form consisting of different generations sharing a common household
kálá páni literally black water, used as a metaphor for the great ocean, or the concept of 'the seven seas'
kantráki contractor, name of a British Indian immigrant
katahar Surinamese name for unripe jackfruit, eaten as a vegetable
kathá story, religious ceremony among Hindus
karhi thick soup of yellow peas
khetihar Hindostani for someone who works the land, usually a small farmer
kitáb book, religious ceremony among Muslims
londá ke nác dance performed by a man dressed as a woman, also called the 'transvestite dance'
masálá curry
mangoes(ten) small predators that feed on other small animals
Muharram Islamic festival of mourning
muran sanskár Hindu christening ceremony during which a child's head is shaved bald
napi sweet potato, a type of yam
neem medicinal leaf of the neem tree, often used by Hindus
paddy/padi unhulled rice
pakorá deep-fried savoury snack made of various vegetables in a batter made of flour
pandit Hindu priest
patjoll working the ground with the patjol (a kind of hoe)
pesi collective name of pulses, including beans
phagwa Hindu festival symbolising various events such as New Year, spring and a rich harvest
photo (pronounced: fôtô) means city, referring to Paramaribo
phulauri savoury lentil snack
podiná kind of mint
Ramlilá play about the life of the Hindu god Rama
Rasgullá syrupy round balls for dessert, made of semolina cooked in sugar syrup
Rasmaláy a juicy 'cheesecake' made from milk
Rexora rice variety
Roti Hindostani dish, similar to an unsweetened 'pancake'
samosá deep-fried triangular pie
sardár supervisor, head, boss

sehená vegetarian snack made of tajer leaf and mung bean
sitár Indian stringed instrument
skrivimankoti rice variety
tablá, tabla drum-shaped percussion instrument
taziyá Muslim procession and play during Muharram
tjatnie (chutney) spicy sambal made from fruit
tulsi basil
urdi mung bean, split peas
zamindar Indian landowner
zebu bullock

References

Aas S (1980) The relevance of Chayanov's Macro Theory to the case of Java. In: Hobsbawn EJ, Kula W, Mitra A et al (eds) Peasants in history: Essays in honour of Daniel Thorner, eds. Oxford University Press, Calcutta, pp 221–248

Abrahamse J (1959) Padi: Verslag van de in 1958 Gehouden Opnamen Betreffende de Oppervlakte en de Produktie. Ministerie van LVV, Paramaribo

Adhin JH (1960) Over de 'joint family' der Hindostanen. Nieuwe West-Indische Gids 40(1):17–27

Adhin JH (1961) Development planning in Surinam in historical perspective. Stenfert Kroese, Leiden

Adhin JH (1969) Ontstaan en Ontwikkeling van de zg. Aziatische Huwelijkswetgeving; overdruk uit: Een Eeuw Surinaamse Codificatie. Paramaribo

Akram-Lodhi H (2001) Vietnam's agriculture: Is there an inverse relationship? Den Haag: ISS Working Papers Series no. 348

Alexander C, Chatterji J, Jalais A (2018) The Bengal diaspora. Rethinking Muslim migration. Routledge, London

Allen R (1999) Slaves, freedman, and indentured laborers in colonial Mauritius. Cambridge University Press, Cambridge

Ansoms A, Verdoodt A, Van Ranst E (2008) The inverse relationship between farm size and productivity in rural Rwanda. Discussion Paper 2008.09, Antwerpen: University of Antwerpen

Arya U (1968) Ritual songs and folksongs of the Hindus of Surinam. Doctoral dissertation, University of Utrecht

Azimullah E, Ganpat H, Lutchman W (1963) Van Brits-Indisch Emigrant tot Burger van Suriname. Wieringa, s-Gravenhage

Ballagh J (2004) White servitude in the colony of Virginia. Study of the system of indentured labor in the American colonies. Kessinger Publishing, Whitefish (Montana)

Banking 1948 Rapport Betreffende het Bankwezen in Suriname. No place mentioned

Bates C (2017) Some thoughts on the representation and misrepresentation of the colonial south Asian labour diaspora. South Asian Stud 33(1):7–22. https://doi.org/10.1080/02666030.2017.1300372

Beck S (1924a) De Verschillende Bedrijfsvormen in den Klein–Landbouw met Betrekking tot Rasaanleg en Volksgewoonten. In: Beck S (ed) Lezingen over Surinaamsche Problemen. Kersten, Paramaribo, pp 1–31

Beck S (1924b) De Creolen en de Klein–Landbouw. In: Beck S (ed) Lezingen over Surinaamsche Problemen. Kersten, Paramaribo, pp 1–34

R. Gowricharn, *The Girmitiya Peasants in Suriname*,
https://doi.org/10.1007/978-3-031-67961-2

Beck S (1924c) Landbouwbedrijfsvorm en landbouwopleiding. In: Beck S (ed) Surinaamsche Problemen. Kersten, Paramaribo, pp 1–44
Becker G (1964) Human capital: A theoretical and empirical analysis with special reference to education. Chicago University Press, Chicago
Beckford G (1972) Persistent poverty. Underdevelopment in plantation economies of the third world. Oxford University Press, New York
Bhagwanbali R (1996) Contracten voor Suriname. Arbeidsmigratie vanuit Brits-Indië onder het indentured-labourstelsel, 1873–1916. Doctoral dissertation, Radboud University (Nijmegen)
Bhagwanbali R (2010) Hindoestaanse migranten onder het indentured labour systeem naar Suriname, 1873–1916. De nieuwe awatar van slavernij. Amrit, Den Haag
Bihari B (1974) Het jajmani-systeem in India en aspekten hiervan onder Hindustanis in Suriname. Varekamp, Paramaribo
Bisnauth D (2000) The settlement of Indians in Guyana 1890–1930. Peepal Tree Press, Leeds
Blumer H (1954) What is wrong with social theory? Am Sociol Rev 19(1):3–10
Blunt A (2003) Collective memory and productive nostalgia: Anglo-Indian home-making at McCluskieganj. Environment and Planning D: Society and Space 21:717–738
Boeke JH (1930) Dualistische Economie. Van Doesburgh, Oratie Leiden
Bosz AJAQ, Albrecht AJ (1954) Drie Eeuwen Grondpolitiek in Suriname. Van Gorcum & Comp, Assen
Bovenkerk F (1975) Emigratie uit Suriname. Antropologisch-Sociologisch Centrum, Amsterdam
Brandon HG (1925) Het Surinamevraagstuk. West-Indische Gids 6(7):481–560
Brandon HG (1933) Proefneming voor Machinale Hoeverijstculuur op de plantage Guadeloupe in Suriname in 1925–1926. West-Indische Gids 14(1):81–98
Brandon HG (1937) De Proefneming van de Surinaamse Hoeve Maatschappij. West-Indische Gids 18(1):293–306
Braverman H (1974) Labor and monopoly capital: The degradation of work in the twentieth century. Monthly Review Press, New York
Budget Survey (1953) Budgetonderzoek gehouden in Suriname in Februari–Maart 1952. Stichting Planbureau, Paramaribo
Bürer T (1956) Het Niet–Gemechaniseerde Rijstbedrijf van 4 ha. De Surinaamse Landbouw 4(5):170–174
Bürer T (1957) Mechanisatie van het Kleine Rijstbedrijf. De Surinaamse Landbouw 5(5/6):196–202
Burger JW (1928) Vergelijkend Overzicht van de Immigratie en Blijvende Vestiging van Javanen en Brits-Indiërs in Suriname. De Economist 77(1):422–445
Caribbean (1955) Caribbean market survey, Rice. Port of Spain, Caribbean Commission
Carney J (2001) Black rice. The African origins of rice cultivation in Africa. Harvard University Press, Cambridge/London
Carrier J (2018) Moral economy. 'What's in a name? Anthropological Theory 18(1):18–35
Carter M (1995) Servants, sirdars and settlers. Indians in Mauritius 1834–1874. Oxford University Press, Delhi
Carter M (2006) Mauritius. In: Lal B, Reeves P, Rai R (eds) The Encyclopedia of the Indian Diaspora. Singapore, Editions Didier Millet, pp 262–272
Carter M (2009) Subaltern networks in a colonial diaspora: A study of Indian migrants and Mauritius. In: Rai R, Reeves P (eds) The South Asian Diaspora. Transnational networks and changing identities. Routledge, London, pp 45–56
Chin H, Buddingh H (1987) Surinam. Politics, economics and society. Francis Printers, London/ NY
Choenni C (2016) Hindostaanse contractarbeiders 1873–1920. Van India naar de plantages in Suriname. LM Publishers, Volendam
Clarke C, Peach C, Vertovec S (eds) (1990) South Asians overseas: Migration and ethnicity. Cambridge University Press, Cambridge
Comins DWD (1892) Note on emigration from the East Indies to Surinam or Dutch Guyana. Printed at the Bengal Secretariat Press, Calcutta

Cornell S, Hartmann D (2007) Ethnicity and race: Making identities in a changing world. Sage, London
Courbois OA (1958) The financing of agriculture in Surinam. Caribbean Commission, Trinidad
De Cock Buning W (1926) Rapport over Suriname, uitgebracht aan de Vereeniging van Kamers van Koophandel en Fabrieken in Nederland
De Haan A (2002) Migration and livelihoods in historical perspective: A case study of Bihar. India Journal of Development Studies 38(5):115–142
de Houven V, van Oordt WJ (1900) Een en ander over kleinen landbouw in de kolonie Suriname. De Economist 49(2):599–615
De Klerk CJM (1941) De Brits-Indiërs in Suriname. Nieuwe West-Indische Gids 24(1):97–117
De Klerk CJM (1953) De Immigratie der Hindostanen. Urbi et Orbi, Amsterdam
De Koning A (1998) Surinaamse Sita's: veranderde invulling van vrouwelijkheid binnen de Hindoestaanse gemeenschap in Suriname vanaf 1950. EUR, faculteit Historische en Kunstwetenschappen, Rotterdam
De Rooij W, Van Schaaijk M (1983) Ekonomische Ontwikkeling van Suriname. Vaco, Paramaribo
De Vos HCP (1938) Rapport omtrent de te nemen maatregelen in het belang van de verbetering van den waterbouwkundigen toestand in Suriname
De Vos HCP (1940) De economische hulpbronnen van Suriname. De Economist 89(1):122–143
De Vos L (1956a) Storage and processing of rice in Suriname. Ministerie van LVV, Paramaribo
De Vos L (1956b) De opslag van padie in Suriname. De Surinaamse Landbouw, jrg 4(5):181–192
De Vries M (1965) De Landbouwers van Suriname. Planbureau, Paramaribo
De Wit TPM (1960) The Wageningen rice project: a study on the development of a mechanized rice farming in the wet Tropics. Doctoral dissertation, Wageningen
Deerr N (1950) A history of sugar, vol 2. Chapman and Hall, London
Despres L (1967) Cultural pluralism and nationalist politics in Guyana. Rand McNally, Chicago
Dew E (1978) The difficult flowering of Surinam; Ethnicity and politics in a plural society. Martinus Nijhoff, The Hague
Dijkhuis LJ (1947) Moderne Landbouw in Suriname. Paris, Amsterdam
Djurfeldt G (1984) Classical discussions of capital and peasantry: A critique. In: Harris J (ed) Rural development: Theories of peasant economy and agrarian change. Hutchinson & Co, London, pp 139–159
DLP (n.d.) Departement Landbouwproefstation, annual reports, several years
DLS (1969) Derde Landbouwtelling Suriname, november 1969. Ministerie van LVV/ABS, Paramaribo
Dorward A (1999) Farm size and productivity in Malawian smallholder agriculture. J Dev Stud 35(5):141–161
DSL (1960) De Surinaamse Landbouw, periodical
Dyer G (1991) Farm size – farm productivity re–examined: Evidence from rural Egypt. J Peasant Stud 19(1):59–92
Easthope H (2004) A place called home. Hous Theory Soc 21(3):128–138
Ehrenburg H, Meyer M (2015) Bouwen aan de Wilde Kust. LM Publishers
Elahi J (2017) Hindostaanse gemeenschapsdansen. In: Gowricharn RS (ed) Onbeschreven Erfgoed. Perspectieven op de Surinaams–Hindostaanse Cultuur. LM Publishers, Volendam
Elahi J (2018) Baithak gáná ke nác: globalisering van een Bhojpuri volksdans. In: Baldewsingh R, Boedhoe N, Mahabier B (eds) Sarnámi Sanskirti. Enkele facetten van de Hindustaanse geschiedenis en cultuur in Suriname en Nederland. Uitgeverij Surinen, Den Haag, pp 199–210
Elenbaas GM (1952) Het Padiebesluit: Memorandum van de Voorzitter van het Surinaams Bedrijfsleven. Vereniging Surinaams Bedrijfsleven, Paramaribo
Eltis D (ed) (2002) Coerced and free migration. Global perspectives. Stanford University Press, Stanford
Emmer P (1984) Een goede keus? De migratie van vrouwelijke contractarbeiders van Brits-Indië naar Suriname. OSO, Tijdschrift voor Surinaamse Taalkunde, Letterkunde en Geschiedenis 3(2):213–234

Emmer P (1985a) De werving van kontraktarbeiders in India voor tewerkstelling in Suriname, 1872–1916. OSO, Tijdschrift voor Surinaamse Taalkunde, Letterkunde en Geschiedenis 4(2):147–157

Emmer P (1985b) The great escape: The migration of female indentured servants from British India to Surinam, 1873–1916. In: Richardson D (ed) Abolition and its aftermath: The historical context, 1790–1916. Frank Cass, London, pp 245–266

Emmer P (1986) The meek Hindu: The recruitment of indentured labourers for service overseas 1870–1916. In: Emmer P (ed) Colonialism and Migration: Indentured Labourers Before and After Slavery. Martinus Nijhoff, Dordrecht, pp 187–207

Emmer P (1987) The position of Indian women in Suriname. Boletin de Estudios Latino americanos y del Caribe 43:115–120

Encyclopaedie (1917/1981) Benjamins H, Snelleman J (eds) Encyclopaedie van Nederlandsch West-Indië. Emmering (facsimile), Amsterdam

Eswin (1956) Jaaruitgave van de Economische Stichting West Indië-Nederland, 's-Gravenhage

Eysvoogel WF (1950) Mechanisatie van de Surinaamse Landbouw. Economisch-Statistische Berichten 1721(3 mei):350–353

Eysvoogel WF, Van Beukering JA, Verhoog JM (1948) Rapport Omtrent de Ontwikkelingsmogelijkheden op Landbouwkundig Gebied in de Westelijke Helft van de Surinaamse Kustvlakte. Wageningen

Fenton S (2003) Ethnicity. Key concepts. Polity Press, Malden

Fokken M (2018) Beyond being koelies and kantráki. Constructing Hindostani identities in Suriname in the era of indenture 1873–1921. Doctoral dissertation, University of Groningen, Groningen

Freund B (1995) Insiders and outsiders. The Indian working class of Durban 1910–1990. Heinemann, Portsmouth NH

Frickers J (1939) Vleeschkeuring, Vleesch en Slachtveeteelt in Suriname. Eben Haezer, Paramaribo

Gaurav S, Mishra S (2015) Farm size and returns to cultivation in India: Revisiting an old debate. Oxf Dev Stud 44(2):165–193

Gautam M (1999) The construction of the Indian image in Surinam: Deconstructing colonial derogatory notions and reconstructing Indian identity. In: Mahin G, Narine D (eds) Sojourners to Settlers: Indian Migrants in the Caribbean and the Americas. Windsor Press, New York, pp 125–181

Geertz C (1971) The integrative revolution: Primordial sentiments and civil politics in the new states. In: Welch C (ed) Political modernization: A reader in comparative political change Belmont. Duxbury Press, CA, pp 198–218

Geertz C (1983/2000) Local knowledge: Further essays in interpretative anthropology. Basic Books, New York

Gillion KL (1962) Fiji's Indian Migrants: A History to the End of Indenture in 1920. Oxford University Press, Melbourne

Glans CL (1963) Nota Betreffende de Rijstpelmolenindustrie in Suriname en Inzake de te Volgen Saneringspolitiek. LVV, Paramaribo

Gobardhan L (2001) Onderwijs Als Sleutel Tot Maatschappelijke Vooruitgang. Een Taal- En Onderwijsgeschiedenis Van Suriname, 1651–1975. Doctoral dissertation, University of Leiden

Goerdayal L (2019) Planten met spirituele en devotionele betekenis voor Hindoes in Suriname en Nederland. In: Baldewsingh R, Boedhoe N, Mahabier B (eds) Sarnámi Sanskirti 2. Enkele facetten van de Hindustaanse geschiedenis en cultuur in Suriname en Nederland. Uitgeverij Surinen, Den Haag, pp 243–264

Gonesh E (2015) Jagernath Lachmon. Verbroederingspolitiek in Suriname. LM Publishers, Utrecht

Goodman D, Redclift M (1981) From peasant to proletarians: Capitalist development and agrarian transitions. Basil Blackwell, Oxford

Gowricharn RS (1985) International trade and the abolition of slavery. Caraïbisch Forum 2(2/3):70–100

Gowricharn RS (1991) Economische transformatie en de staat. Over agrarische modernisering en economische ontwikkeling in Suriname, 1930–1960. Doctoral dissertation, University of Utrecht

Gowricharn RS (2013) Ethnogenesis: The case of British Indians in the Caribbean. Comp Stud Soc Hist 55(2):388–418

Gowricharn RS (2016) Hindostaanse heritage in Suriname. OSO, tijdschrift voor Surinaamse Letterkunde en geschiedenis, November, 115–129

Gowricharn RS (2019) Het spook van de koelie: de stadshindustaan. In: Baldewsingh R, Boedhoe R, Mahabier B (eds) Sarnámi Sanskirti. Enkele facetten van de Hindustaanse geschiedenis en cultuur in Suriname en Nederland, deel 2. Surinen, Den Haag, pp 107–122

Gowricharn RS (2020) Multiple homemaking. The ethnic condition in the Indian diaspora. Routledge, London

Gowricharn RS (2023) Beyond the Kala Pani. Indenture Papers Studies on Girmitiyas 3:25–38. https://girmit.ac.fj/wp-content/uploads/ip-2023/gowricharn.pdf

Grierson GA (1883) Report on colonial emigration from the Bengal presidency. Government of Bengal, Bengal

Grierson GA (1885/1975) Bihar peasant life, being a discursive catalogue of the surroundings of the people of that province, with many illustrations from photographs taken by the author. Cosmo Publications, Delhi

Grondslagen (1952) De Grondslagen van het Tienjarenplan voor Suriname. Paramaribo, Den Haag

Hale H (2004) Explaining ethnicity. Comp Pol Stud 37(4):458–485

Haraksingh K (1999) Indentured and self-emancipation. In: Lamming G (ed) Enterprises of the Indies. Institute of the West Indies, Tunapuna, Trinidad & Tobago, pp 38–42

Hassankhan M (1993) De Immigratie en haar gevolgen voor de Surinaamse samenleving. In: Gobardhan-Rambocus L, Hassankhan M (eds) Immigratie en Ontwikkeling. Emancipatie van Contractanten. Anton de Kom University, Paramaribo, pp 11–35

Hassankhan M (2013) Kahe Gaile Bides – Why did you go overseas? An introduction in emotional aspects of migration history: A diaspora perspective. In: Mahabir K (ed) Caribbean issues in Indian diaspora. Serial Publications, New Delhi, pp 3–35

Hassankhan M (2016) Islam and Indian Muslims in Suriname: A struggle for survival. In: Goolam V, Roopnarine L (eds) Indentured Muslims in the diaspora. Identity and belonging of minority groups in plural societies. Manohar, New Delhi, pp 183–228

Hasselbach O (1966) The introduction of lodging resistant rice varieties in peasant agriculture. De Surinaamse Landbouw 14:45–55

Heilbron W (1982) Kleine boeren in de schaduw van de plantage. De politieke ekonomie van de na-slavernijperiode in Suriname. Doctoral dissertation, Erasmus University Rotterdam

Heilbron W, Willemsen G (1977) Transformaties in een plantage ekonomie. De politieke ekonomie van Suriname 1863–1900. Self published

Heilbron W, Willemsen G (1980a) Goud- en balata-exploitatie in Suriname: Nieuwe productiesectoren en nieuwe vormen van afhankelijkheid na 1863. Caraïbisch Forum 1(1):66–84

Heilbron W, Willemsen G (1980b) Goud- en balata-exploitatie in Suriname: Nieuwe productiesectoren en nieuwe vormen van afhankelijkheid na 1863. Caraïbisch Forum 1(1):87–101

Hendriks JAH, Joosten WJM (1954) De Resultaten van een landbouw–proeftelling in de omgeving van Lelydorp. De Surinaamse Landbouw 2(1):21–35

Hewitt WE (1922) Nickerie als landbouwdistrict. West-Indië, Landbouwkundig Tijdschrift voor Suriname en Curaçao 7(4):5–20

Hewitt WE (1941) De sociaal-economische toestand van klein-landbouwers aan de Boven-Commewijne. Jaarverslag van het Departement van Landbouw-Economische Zaken, Vol. not reported, p 73–87

Heybroek F (1978) Evaluatie en Inventarisatie van het project Santo Boma. Ministerie van LVV, Dienst Landbouw–Economische Zaken, Paramaribo

Hindori D (1998) Geschiedenis van onze landbouw. In: Mitrasingh B, Harpal M (eds) Hindostanen. Van contractarbeiders tot Surinamers. Stichting Hindostaanse Immigratie, Paramaribo, pp 19–39

Hira S (1982) Van Priari tot en met De Kom: de geschiedenis van het verzet in Suriname 1630–1940. Futile, Rotterdam

Hira S (2000) Terug naar Uttar Pradesh. Op zoek naar de wortels van Surinaamse Hindostanen. Amrit, Den Haag

Hobsbawm E, Ranger T (eds). (1983/1992). The invention of tradition. Cambridge University Press, Cambridge

Hoefte R (1987a) Female indentured labor in Suriname: For better or for worse? Boletín de Estudios Latinoamericanos y del Caribe 42:55–70

Hoefte R (1987b) The position of female British Indian and Javanese contract laborers in Suriname. A last word. Boletín de Estudios Latinoamericanos y del Caribe 43:121–123

Hoefte R (1990) De betovering verbroken. De migratie van Javanen naar Suriname en het Rapport Van Vleuten 1909. Foris Publications, Dordrecht

Hoefte R (1998) In place of slavery. A social history of British Indian and Javanese laborers in Suriname. University Press of Florida, Gainesville

Hoppe R (1976) Het politiek systeem van Suriname: elite–kartel demokratie. Acta Politica 2:145–177

Introductory (1953) Introductory report on mechanizing agriculture in Suriname, Report no. 1. Wageningen/Holland, Netherlands Advisory Board for the Mechanization of Agriculture in the Overseas Territories

Jayawardena C (1963) Conflict and solidarity in a Guyanese plantation. The Athlone Press, London

Junker L (1940) Het irrigatie vraagstuk in Suriname. West-Indische Gids dl 23(22):226–233

K.v.K (1916) Jaarverslag van de Kamer van Koophandel en Fabrieken. Paramaribo

K.v.K. (1950) Kamer van Koophandel (Chamber of Commerce)

Karsten R (1930) De Brits–Indiërs in Suriname. Een korte schets benevens een handleiding voor de beginselen van het Hindi. Martinus Nijhoff

Kasteleyn JSC, Kluvers BJ (1919–1920). Resumé van het rapport (1919) der Studie-Commissie van het Suriname-Studie-Syndicaat. De West-Indische Gids 110:287–316

Khan MR (2005) An autobiography of an Indian indentured labourer. Amrit, Den Haag

Kielstra JC (1918) Recht en Werkelijkheid. Inaugurele rede, Landbouw Hogeschool Wageningen

Kielstra JC (1927) Nieuwe mogelijkheden voor Suriname. De Economist 76(1):181–196

Kielstra JC (1946) Application of Netherlands Indies agrarian principles to Surinam (Netherlands Guyana). In: Caribbean land tenure symposium, Washington, pp 335–347

Klass M (1961) East Indians in Trinidad. A study of cultural persistence. Waveland Press, Long Grove

Klein WC (1946) Opmerkingen over Industrie in Suriname. Eswin 1:3–12

Kool R (1955) De toekomstige binnenlandse afzet van de Surinaamse landbouw. De Surinaamse Landbouw 3(2):48–53

Kool R (1964) Agricultural planning in Suriname 1950–1960: An evaluation. Wageningen, Veenman & Zonen

Kraijo M (2022) Destined to leave Hindustan for Suriname? Explaining the motivation behind repatriation and settlement of Hindustani labour migrants in Suriname, 1873-1940. Tijdschrift voor Sociaal-Economische Geschiedenis (TSEG) 9(3):37–68. https://doi.org/10.52024/tseg.10894

Kruijer G (1949) Landbouw-Cooperatie onder de Aziatische bevolkingsgroepen van Suriname en Brits-Guyana. Nieuwe West-Indische Gids 30(1):209–235

Kruijer G (1951) Urbanisme in Suriname. Tijdschrift van het Koninklijk Nederlands Aardrijkskundig Genootschap, jrg 58:31–63

Kruijer G (1973) Suriname, Neokolonie in Rijksverband. Boom, Amsterdam

Kumar A (2016) Feeding the Girmitiya 1: Food and drink on indentured ships to the sugar colonies. Gastronomica: The Journal of Critical Food Studies 16(1):40–51

Kumar A (2017) Coolies of the Empire. Indentured Indians in the sugar colonies 1830–1920. Cambridge University Press, Cambridge

Lal B (1983a) Girmitiyas. The origins of the Fiji Indians. Australian National University Press, Canberra

Lal B (1983b) Indian indentured historiography: A note on problems, sources and methods. Pacific Studies 6(2):33–50

Lal B (2000) Chalo Jahaji. On a journey through indenture in Fiji. Australian National University Press, Canberra

Lal B (2006) Fiji. In: Lal B, Reeves P, Rai R (eds) The Encyclopedia of the Indian diaspora. Singapore, Editions Didier Millet, pp 370–382

Lalmohamed B (1992) Hindostaanse vrouwen: de geschiedenis van zes generaties. Van Arkel, Utrecht

Lampe PHJ (1927) Suriname. Sociaal–Hygiënische Beschouwingen. De Bussy, Amsterdam

Lamur H (1973) The demographic evolution of Surinam 1920–1970: A socio-demographic analysis. Martinus Nijhoff, The Hague

Laurence K (1994) A question of labour: Indentured immigration into Trinidad and British Guiana 1875–1917. Ian Randle Publishers, Kingston

Legêne P (1936) Het volk van het groote heimwee. Zeist, EBG

Lewis WA (1950) The industrialisation of the British West Indies. Caribbean Economic Review 2(1):1–61

Lewis WA (1954) Economic development with unlimited supplies of labour. In: Agarwala AN, Singh SP (eds) The economics of underdevelopment. Oxford University Press, London, pp 400–449

Leys JJ (1916a) Het bemesten van onze rijstvelden. West-Indië, Landbouwkundig Tijdschrift voor Suriname 1(1):1–11

Leys JJ (1916b) Het bemesten van onze rijstvelden. West-Indië, Landbouwkundig Tijdschrift voor Suriname 1(2):1–8

Leys JJ (1917) Onze veeteelt. West-Indie, landbouwkundig tijdschrift voor Suriname en Curaçao 2(4):105–111

Leys JJ (1920a) Het Arbeidersvraagstuk. West-Indie 3:114–121

Leys JJ (1920b) Onze Veestapel. West-Indië, landbouwkundig tijdschrift voor Suriname en Curaçao 2(2):69–75

Leys JJ (1921) De Kleine Landbouwers en het Klein Landbouwbedrijf. West-Indië, Landbouwkundig Tijdschrift voor Suriname en Curaçao 6(1):31–44

Leys JJ (1925a) De veeteelt in Suriname. Nieuwe West-Indische Gids 6(1):405–418

Leys JJ (1925b) De veeteelt in Suriname. Nieuwe West-Indische Gids 6(1):441–460

LEZ (1941) Jaarverslag van het Department van Landbouw–Economische Zaken. Several years

Liems JA (1918) Is de Creool voorbestemd om dienstbaar te worden aan den Brits–Indiërs? West-Indië, Landbouwkundig Tijdschrift voor Suriname en Curaçao 3(3):140–144

Liems JA (1921) Toepassing van machinale bewerking in Suriname. West-Indië 2:10–14

Liems JA (1923) Het klein-landbouwbedrijf en de vruchtencultuur in Suriname. West-Indische Gids dl. 5(4):625–638

Liems JA (1926) Over den kleinen landbouw in Suriname. West-Indische Gids 7(1):371–384

LPS (1924/1927) Jaarverslagen van het Landbouwproefstation Suriname. Agricultural Experimental Station, Published irregular

Luning H (1966a) Bedrijfsgrootte, werkgelegenheid en agrarische planning. De Surinaamse Landbouw 14(1&2):36–44

Luning H (1966b) Enkele kanttekeningen aangaande de taak van de landbouw bij de ekonomische ontwikkeling van Suriname. De Surinaamse Landbouw 14(4):117–135

Luning H (1967) Economic aspects of low labour-income farming. Doctoral dissertation, University Wageningen

Luning H (1968) Transforming traditional rice farming: The Surinamese experience. Neth J Agric Sci 16:253–263

Luning H, Sital P (1979) The economic transformation of small holder rice farming in Surinam. In: Cross M, Marks A (eds) Peasants, plantations and rural communities in the Caribbean. Leiden, KITLV, pp 193–219

Mahabir K (2004) Animals from India in the Caribbean. Zoological survey of India 103(Parts 1–2). Government of India, Kolkata, p 179–183

Mahabir K (2008) Medicinal and edible plants used by East Indians of Trinidad & Tobago. Chakra Publishing House, Madison

Majumder M (ed.) (2010) Kahe Gaile Bides (Why did you go overseas?). In: On Bhojpuri migration since the 1879's and contemporary culture in Uttar Pradesh and Bihar, Suriname and the Netherlands. Mango

Mandle J (1973) The plantation economy: Population and economic change in Guyana, 1838–1960. Temple University Press, Philadelphia

Marhé RM (1985) Sarnami Byäkaran: een elementaire grammatica van het Sarnami. Stichting voor Surinamers voor Den Haag, Den Haag

Marx K (1867) Capital. A critique of political economy, vol 1. Penguin Books, Middlesex

Mayer A (1966) Quasi-groups in the study of complex societies. In: Banton M (ed) The social anthropology of complex societies. Tavistock Publications, London, pp 97–122

Mayer A (1973) Peasants in the Pacific. A study of Fijian Indian rural society. Routledge & Kegan Paul, London

McNeill J, Lal C (1915) Report to the government of India on the conditions of Indian immigrants in four British colonies and Surinam, Part II. London

Meel P (1994) Verbroederingspolitiek en nationalisme: het dekolonisatievraagstuk in de Surinaamse politiek. Bijdragen en Mededelingen betreffende de geschiedenis der Nederlanden 109(4):638–659

Memorie (1919) Memorie van het Comité tot Bevordering van den Middenstandslandbouw in Suriname, op 29 augustus 1919 den Gouverneur Aangeboden. West-Indische Gids 1(2):163–176

Menke J, Sno I (2016) Ras en etniciteit in volkstellingen van Suriname. In: Mencke J, Sno I (eds) Mozaïek van het Surinaamse volk. Volkstellingen in demografisch, economisch en sociaal perspectief. UvS/ABS, Paramaribo, pp 76–94

Menkman WR (1940) Landbouw-economische politiek in Suriname. West-Indische Gids 23(22):12–19

Mhango B (1980) Gap analysis applied to the Surinamese economy. Doctoral dissertation, University van Suriname

Mintz S (1973) A note on the definition of peasantries. J Peasant Stud 1(1):91–106

Mintz S (1983) Reflections on Caribbean peasantries. New West Indian guide 57(1–2):1–17

Mishra AK (2015) Sardars, Kanganies and Maistries: Intermediaries in the Indian Labour Diaspora during the Colonial Period. In: Wadauer S, Buchner TH, Mejstrik A (eds) The History of Labour Intermediation: Institutions and Finding Employment in the Nineteenth and Early Twentieth Centuries. Berghahn Books, New York, Oxford, pp 368–387. https://doi.org/10.1515/9781782385516-014

Mitrasing F (1959) Tien jaar Suriname. Van afhankelijkheid tot gelijkgerechtigdheid 1945–1955. Doctoral dissertation, University of Leiden

Mohan P (2007) Jahajin. HarperCollins, Delhi

Mohapatra P (1995) Restoring the family: Wife murders and the making of sexual contract of Indian immigrant labour in British Caribbean colonies 1860–1920. Stud Hist 11(2):227–260

Morenc J (1988) Surinaamse Kleine Landbouw en Landbouwbeleid: Een Structurele Analyse. Doctoral dissertation, Radboud University of Nijmegen

Mügge L (2012) Ideologies of nationhood in sending-state transnationalism: comparing Surinam and Turkey. Ethnicities 13:338

Munro D (2014) Conclusion: On resistance and accommodation. In: Lal B, Munro D (eds) Hassankhan M. Resistance and Indian experiences. Comparative perspectives, Manhor, pp 293–311

Narayan B (2017) Culture and the emotional economy of migration. Routledge, London

Nath D (1950/1970) A history of Indians in Guyana. Second revised edition published by the author, London
Nayar P (2015) The postcolonial studies dictionary. Wiley Blackwell, Oxford
NEI (1952) Onderzoek Naar de Afzetmogelijkheden van Surinaamse Rijst. Nederlands Economisch Instituut, Rotterdam
Northrup D (1995) Indentured labor in the age of Imperialism, 1834–1922. Cambridge University Press, Cambridge
Ontwikkeling (1956) De sociaal-economische omstandigheden in Suriname, in het byzonder betreffende de wenselijkheid tot versnelde ontwikkeling van industrie en nijverheid. Departement van Economische Zaken, Paramaribo
Ooft C (1972) Ontwikkeling van het Constitutionele Recht van Suriname. Doctoral dissertation, University of Leiden
Panday N (1959) Agriculture in Surinam 1650–1950: An inquiry into the causes of its decline. Paris, Amsterdam
Patnaik U (1976) Class differentiation within the peasantry: an approach to analysis of Indian Agriculture. Econ Polit Wkly 39:82–101
Premdas R (1996) Race and ethnic relations in Burnhamite Guyana. In: Dabydeen D, Samaroo B (eds) Across the dark waters: Ethnicity and Indian identity in the Caribbean. Macmillan, London and Basingstoke, pp 39–64
Price R (2007) Some anthropological musings on creolization. Journal of Pidgin and Creole Languages 22(1):17–36
Pyttersen T (1922) Rapport omtrent de uitkomsten van een voorloopig onderzoek naar de levensvatbaarheid van landbouwersbedrijven in Suriname met toepassing van moderne cultuurmethodes. s–Landsdrukkerij, s-Gravenhage
Raghoenandan U (1994) Etnobotanisch onderzoek bij de Hindostaanse bevolkingsgroep in Suriname. UvS/BBS (Nationaal Herbarium van Suriname)
Ramadhin H (2014) De historische ontwikkeling van de rijstbouw. Het verleden, het heden en de toekomst. Published in-House
Ramdin R (2000) Arising from bondage: A history of Indo-Caribbean people. Tauris Publishers, London/ New York
Ramsoedh H (1990) Suriname 1933–1944: Koloniale politiek en beleid onder gouverneur Kielstra. Doctoral dissertation, University of Utrecht
Ramsoedh H (2018) Surinaams Onbehagen. Een sociale en politieke geschiedenis van Suriname 1865–2015. Verloren, Hilversum
Rapport (1920) Rapport van de Studie Commissie van het Suriname Studie Syndicaat (1920). Gebr. Tuinzin, Rotterdam
Rapport (1945) Rapport van de Commissie ter bestudeering van de na-oorlogse problemen in Suriname, Paramaribo
Reddy M (2016) Social movement and the Indian diaspora. Routledge, London/New York
Ref. (1921) Landbouwvoorlichting aan de Britsch-Indiërs. West–Indië, Landbouwkundig Tijdschrift voor Suriname en Curaçao 6(2):33–36
Rekest (1926) Rekest van de Ondernemersraad aan den Minister van Koloniën d.d. juli 1926. Helder
Relyveld WL (1954) Mechanisatie van de landbouw in Nickerie. De Surinaamse Landbouw 2(2):75–84
Rest CPM (1974) Het departement van Landbouw, Veeteelt en Visserij. Onderzoeksproject Sociale Ontwikkelingsstrategie. University of Amsterdam, Amsterdam
Roopnarine L (2003) East-Indian indentured emigration to the Caribbean: Beyond the push and pull model. Caribb Stud 31(2):97–134
Roopnarine L (2007) Indo-Caribbean indenture. Resistance and accommodation, 1838–1920. University of the West Indies Press, Mona
Roy P (1981) Transition in agriculture: empirical indicators and results (evidence from Punjab, India). J Peasant Stud 8(2):213–241
Roy T (2018) Inequality in colonial India. Economic History Working Papers, no. 286. The London School of Economic and Political Science

Samaroo B (2006) The Mahatma in the Caribbean. In: Gowricharn R (ed) Caribbean transnationalism. Migration, pluralization, and social cohesion. Lexington Books, Lanham/Boulder, pp 139–153

Samaroo B (2017) Changing Caribbean geographies: Connections in flora, fauna and patterns of settlement from Indian inheritances. Unpublished manuscript, History Department, University of the West Indies, St. Augustine, Trinidad

Schets (1951) Eerste Schets van de Ontwikkelingsmogelijkheden van Suriname, deel ll, De Plannen. Paramaribo/Den Haag

Schiller N, Salazar N (2013) Regimes of mobility across the globe. J Ethn Migr Stud 39(2):183–200

Scholtens B (1986) Opkomende Arbeidersbeweging in Suriname. Masusa, Nijmegen

Schütz HLH (1941) De Ontwikkeling van den Landbouw in Nickerie gedurende de laatste 10 jaar, 88–100. Jaarverslag van het Departement van Landbouw-Economische Zaken, Paramaribo

Sedney J (2017) De Toekomst van ons Verleden. Democratie, Etniciteit en politieke machtsvorming in Suriname. Vaco, Paramaribo

Seecharan C (2011) Mother India's shadow over El Dorado: Indo-Guyanese politics and identity 1890s–1930s. Ian Randle Publishers, Kingston en Miami

Seigel W (2017) The great leveler. violence and the history of inequality from the stone age to the twenty-first century. Princeton University Press, Princeton

Service E (n.d.) Peasant societies. Encyclopedia Brittanica, Primitive culture – peasant societies, subsistence, rituals | Britannica

Shanin T (ed) (1984) Peasants and peasant societies. Penguin, Harmondsworth

Sheller M (2014) The new mobility paradigm for a live sociology. Current Sociology Review 63(6):789–811

Shepherd V (1993) Transients to settlers: The experience of Indians in Jamaica 1845–1950. Peepal Tree Books, Leeds

Shils E (1957) Primordial, personal, sacred and civil ties. Br J Sociol 8(2):130–145

Singelenberg R (1983) Guyanese gastarbeid in Suriname. Instituut voor Culturele Antropologie, Utrecht

Singh N (2016) My life my story: Narratives of women left behind. In: Hassankhan M, Roopnarine L, Mahase R (eds) Social and cultural dimensions of Indian indentured labour and its diaspora: past and present. New Delhi, Manohar, pp 25–50

Sital JT (2002) How Suriname came to its own green revolution. A study of change in small-holder farming in the Nickerie district, 1900–1986. University of Wageningen, Wageningen

Siwpersad J (1979) De Nederlandse regering en de afschaffing van Surinaamse slavernij (1833–1863). Doctoral dissertation, University of Groningen

Snellen E (1933) De Aanvoer van arbeiders voor den landbouw in Suriname. Doctoral dissertation, University of Wageningen

Sollors W (ed) (1989) The invention of ethnicity. Oxford University Press, New York/Oxford

Speckmann J (1965) Marriage and kinship among the Indians in Surinam. Van Gorcum, Assen

Staal J (1919) Memorie van het Comité tot bevordering van den middenstandslandbouw in Suriname, op 29 augustus 1919 den Gouverneur aangeboden. Overgenomen uit De Surinamer van 2 en 6 november 1919

Stahel G (1933) De Rijstcultuur in Suriname; Amsterdam, z.j. (overdruk uit De Indische Mercuur, 22 en 29 november, en 6 dec 1933), publication of the Landbouwproefstation te Paramaribo, no. 2

Stiglitz J, Walsh C (2006) Principles of Micro-Economics. Norton & Company, New York/London

Surinaamsch Verslag (1950). Jaarverslag van het koloniaal Bestuur ten behoeve van de Nederlandse Regering, annual reports

Surinam (1952) Surinam, Recommendations for a ten-year development program. John Hopkins Press, Baltimore

Tinker H (1974) A new system of slavery: The export of Indian labour overseas 1830–1920. Oxford University Press, London/New York

TJP (1954) Tienjarenplan voor Suriname. Herziene opstelling door de Stichting Planbureau Suriname. Planbureau, Paramaribo
TLS (1959) Tweede Landbouwtelling Suriname, november 1959. Ministerie van LVV, in samenwerking met ABS, Paramaribo
Udehn L (2002) The changing face of methodological individualism. Annu Rev Sociol 28:479–507
Uitbuiting (1970) De Uitbuiting van Suriname. SUN, Nijmegen
Van Andel T, Ruysschaert S (2014) Medicinale en rituele planten van Suriname. LM Publishers/VACO, Arnhem/Paramaribo
Van As JB, Del Prado H (1960) Nota Inzake Garantieprijzen voor Landbouwproducten. Ministerie van LVV, Paramaribo
Van Beukering JA (1952) Types of farming. Caribbean Economic Review IV(1&2):140–163
Van Blom D (1919) Kolonisatie met Nederlandsche Boeren in Suriname. West-Indische Gids 2(1):345–359
Van de Walle J (1945) Suriname, Rapport uitgebracht door het hoofd van den Gouvernements Persdienst in Suriname. Paramaribo
Van de Walle J (1975) Een oog boven Paramaribo: Herinneringen. Querido, Amsterdam
Van den Helm PWHG (1926) Het Suriname vraagstuk. Rapport op verzoek der Regeering uitgebracht
Van Dijk HN (1940) Mechanische Rijstcultuur: Economische en Technische Resultaten Na Zes Jaar Pioniersarbeid te Nieuw–Nickerie, Suriname. Amsterdam: Paris
Van Dijk HN (1952) Suriname's rijstcultuur wordt bedreigd: Open brief aan de Landsminister te Paramaribo. Paramaribo, Nationale Vereniging 'Nickerie Vooruit'. VSB, Paramaribo
Van Drimmelen C (1918) Een Middenstandslandbouw voor Suriname. Uitgave van de R.K. Sociale Cursus, Paramaribo
Van Dusseldorp DBWM (1963a) Het Centraplan voor Suriname: Some aspects of town and country planning in a developing country. Wageningen, Veenman & Zonen
Van Dusseldorp DBWM (1963b) Geografische Mobiliteit en de ontwikkeling van Suriname. Bijdragen tot de Taal-, Land-, en Volkenkunde 199(1):18–55
Van Dusseldorp DBWM (1967) Meerdimensionale Overheidsplanning. De overheidsplanning van Suriname in de periode 1952–1964, getoetst aan een relatiemodel. Doctoral dissertation, Wageningen
Van Gelder P (1984) Werken Onder een Boom; Dynamiek en Informele Sector: de Situatie in Groot-Paramaribo, Suriname. Doctoral dissertation, University of Amsterdam, Amsterdam
Van Kempen MHG (2002) Een geschiedenis van de Surinaamse literatuur. Uitgeverij Okopipi, Paramaribo, pp 279–305
Van Lier R (1977) Samenleving in een grensgebied: Een sociaal–historische studie van Suriname. Emmering, Amsterdam
Van Praag S (1877) Suriname: geen immigratie–leening, maar de beschaving harer negerbevolking, het krachtigste redmiddel dier kolonie: redevoeringen, gehouden in de zitting der koloniale staten van Suriname, van den 12 Februari 1877. Oliver, Amsterdam
Van Stipriaan A (1993) Surinaams Contrast. Roofbouw en overleven in een Caraïbische plantage–economie, 1750–1863. Brill, Leiden
Van Traa A (1946) Suriname 1900–1940. Van Hoeve, Deventer
Van Waesberge R (1976) Suriname, Een stukje derde wereld. Novib, Den Haag
Van Wengen GD (1972) De Javanen in de Surinaamse Samenleving. Sticusa, Amsterdam
Van Wouw JJ (1949) Het Departement van Landbouw-Economische Zaken in Suriname in 1945. Nieuwe West-Indische Gids 30(1):244–249
Verkade-Cartier van Dissel EF (1937) De mogelijkheid van landbouwkolonisatie van blanken in Suriname. Doctoral dissertation, University of Utrecht
VHP (2019) 70 jaar VHP 1949–2019. Zelf Sterk, Samen Sterker. Published by VHP, Paramaribo
Vroon IJ (1963) Voorgeschiedenis, opzet en resultaten van het Surinaamse tienjarenplan. Nieuwe West-Indische Gids 43(1):25–74
Weber M (1904) The Protestant ethic and the spirit of capitalism. Allen & Unwin, London

Weber M (1978) In: Roth G, Wittich C (eds) Economy and society: An outline of interpretative sociology, vol 2. University of California Press, Berkeley

Welvaartsfonds (1955) Samenvattend Eindverslag van het Welvaartsfonds Suriname. Tevens Verslag over het Jaar 1954, Paramaribo

Wenzel T (1908) Schetsen uit den Zendingsarbeid der Evangelische Broedergemeente onder de Brits–Indiërs in Suriname. Vonk & Co

Wesenhagen AC (1886) De middelen door de Staten–Generaal aangegeven tot behoud van der kolonie Suriname van meer nabij beschouwd. Nijgh & Van Ditmar, Rotterdam

Westra P (1921) Contractimmigratie van Britisch–Indiërs. West-Indië, Landbouwkundig Tijdschrift voor Suriname en Curaçao 6(2):3–8

Willemsen G (1980) Koloniale politiek en transformatieprocessen in een plantage-economie. Suriname 1873–1940. Doctoral dissertation, Erasmus University of Rotterdam

Yang A (1979) Peasants on the move: a study of internal migration in India. J Interdiscip Hist 10(1):37–58

Zielhuis L, Girdhari D (1973) Migratie uit Suriname. Afdeling Wetenschappelijk Onderzoek en Planning, Paramaribo

Name Index

R. Gowricharn, *The Girmitiya Peasants in Suriname*,
https://doi.org/10.1007/978-3-031-67961-2

Subject Index

R. Gowricharn, *The Girmitiya Peasants in Suriname*,
https://doi.org/10.1007/978-3-031-67961-2

SPRINGER NATURE

GPSR Compliance

The European Union's (EU) General Product Safety Regulation (GPSR) is a set of rules that requires consumer products to be safe and our obligations to ensure this.

If you have any concerns about our products, you can contact us on ProductSafety@springernature.com

In case Publisher is established outside the EU, the EU authorized representative is:

Springer Nature Customer Service Center GmbH
Europaplatz 3
69115 Heidelberg, Germany